Study Guide

Introductory Chemistry

A Foundation

SIXTH EDITION

Zumdahl/DeCoste

Donald J. DeCoste

University of Illinois at Urbana-Champaign

HOUGHTON MIFFLIN COMPANY BOSTON NEW YORK

Vice President and Publisher: Charles Hartford
Senior Marketing Manager: Laura McGinn
Marketing Assistant: Kris Bishop
Development Editors: Bess Deck, Rebecca Berady Schwartz, Kate Heinle
Editorial Associate: Chip Cheek
Senior Project Editor: Cathy Labresh Brooks
Editorial Assistan: Katherine Roz
Production Coordinator: Sean McGann
Ancillary Producer: Brett Pasinella

Printed in the U.S.A.

ISBN-10: 0-618-80333-5
ISBN-13: 978-0-618-80333-0

23456789-POO-11 10 09 08 07

Contents

Preface

You are probably asking yourself, "What things can I do to help me learn chemistry?" This is a question many students ask when they first begin a chemistry course. The complete answer to this question is not simple, but there are some things you can do to help make your study of chemistry successful.

First, read the textbook assignment at least twice, once before you go to class (so that you will receive maximum benefit from each lecture) and once after going to class (so that you can slowly and carefully review the main points of the lecture). Second, keep up with the homework assignments. Don't wait until the day before the exam to start working the problems. Solving chemistry problems takes time, thought, and practice. You will need to work steadily to become a good problem solver.

This *Study Guide* can provide additional help. The **Chapter Discussion** summarizes important concepts from the chapter. The **Chapter Discussion** is not a rewrite of the chapter nor is it all-inclusive. However, it provides additional explanations for concepts that students generally find particularly difficult.

The **Learning Review** provides problems in addition to those in the text. Each one has a worked-out solution in **Answers to Learning Review**. Try to work these problems first, and then check your answer. Looking at the solutions before you work the problem will not benefit you. A main goal of studying chemistry is to become better at problem solving. Looking at the solutions and understanding them is not the same thing as coming up with the solutions in the first place.

Take advantage of the resources available to you: your instructor, your textbook, this *Study Guide*, and your own personal effort and abilities, and you will get the most out of your chemistry course. Have a good semester!

D.J.D.

CHAPTER 1

Chemistry: An Introduction

INTRODUCTION

Your study of chemistry will require work on your part. Use all of the resources available to you so that you can get the most out of the effort you put into learning chemistry. The textbook should be your primary resource, but do not hesitate to turn to this *Study Guide* for additional help.

This chapter introduces you to how scientists solve problems. Learning how to solve problems is an important part of any chemistry course. Problem solving means more than just calculating a numerical answer. It also includes sifting through the given information, deciding which pieces of information are useful, and finally selecting an approach that will solve the problem. The problem-solving skills you develop can be useful to you throughout your life.

STUDYING CHEMISTRY

What does it mean to "know chemistry"? And what makes chemistry courses among the most difficult courses you take? We have asked these questions of students at the University of Illinois at the beginning of the academic year for the past several years, and while not all of the answers are the same, the spirit of the answers is quite consistent. According to many of the students, "knowing chemistry" essentially means knowing "a bunch of stuff." Generally this "stuff" is said to consist of unrelated, and often counterintuitive, facts, equations, and constants. "Knowing chemistry" also means solving problems. These problems are usually math based and are solved by plugging the right numbers into the right equations. These understandings (or rather misunderstandings) of chemistry may help to explain some of the difficulty with chemistry courses. It is a hard task to memorize a great deal of disparate facts, especially when they seem to go against our original ideas.

So what can be done about this? Can studying chemistry be made more enjoyable? Yes. However, the way of doing this may seem ironic, at least initially. One of the reasons that chemistry is so difficult is because of the way students try to make it easier. What does this mean? Read this passage and see to what extent it sounds like you:

> You attend all the lectures and studiously take notes on everything the professor writes on the board. You become distracted when the professor seems to go on a tangent or derives an equation or proves a point. You dislike it when the professor follows a line of thought that proves to be incorrect just to make a point. You rarely ask questions either of the professor or of yourself during the lecture. You also view others' questions during the lectures as diversions.
>
> You generally do all of the assigned homework questions. You usually go straight to the problems and try to do them with minimal reading of the text. This usually consists of trying a problem, and, if you have difficulty, going to a section in the chapter looking for an example problem. When you do read the text, you generally pay little attention to the graphs and figures and rarely write questions about what you have read. Sometimes you do the problems with the *Solutions Manual* open, but you take only a small peek because once you see how to set up the problem you can usually solve it. You can answer most of the "what" questions, but have difficulty with the "why" questions. You have become adept at using units to solve problems you really do not understand.

Obviously this does not apply to you exactly, but how many of these characteristics come close to describing you? While no one student is exactly like the student described, all of the students we have talked with have some of these characteristics. And this is not to say that these ideas are completely unreasonable; in fact, many of the qualities are admirable. For example, the described student goes to all the lectures and does all the homework. In fact, some students as described above do well in their chemistry courses. So what is the problem?

While some of the students like the one described are successful in their chemistry courses, success often comes with much unnecessary frustration. Worse yet, many of the students described have so much difficulty they either drop the course or end up doing quite poorly. Many of the latter students spend more and more time studying while getting less and less out of it.

What is most ironic about this is there is a cyclical pattern that we have seen for good, hardworking students who end up doing poorly. They see chemistry as unrelated facts, and so they study for it that way. They expect the material to be counterintuitive, so it is. They do not think of chemistry as being about concepts, so it is not. By using shortcuts, mnemonic devices, and memorized algorithms, they actually make chemistry more difficult because when they come across a problem they have not seen before (which occurs in chemistry courses), they have no idea what to do with it. They are not used to trial and error and are unfamiliar with the idea of making mistakes as a way of learning. Thus they learn less. The more difficult the material becomes, the more they study (assuming they do not give up). But they do not know how to study more effectively.

When you understand chemistry, you can solve unique problems by applying fundamental ideas, not by using memorized solutions or plugging numbers into an equation. You should also understand the following:

1. **Chemistry content should make sense.** Many of our students have stated that when answering a question they think about what makes sense and then say the opposite. Because, after all, this is chemistry, and "the answer is always the opposite of what you think it should be." While the students say this somewhat in jest, many actually believe it to some extent. One of the goals of learning chemistry is to be able to understand, explain, and predict real phenomena in the real world. You should strive for an understanding of the connectedness of the ideas; you should not just memorize equations and methods for solving problems. The ideas and concepts support each other and are consistent. Make sure to understand it this way.

2. **Explaining is different from knowing or remembering, and formulas and equations are not explanatory devices.** For example, claiming "elements want to have an electron configuration like noble gases" may help you figure out that oxygen has a 2− charge as an ion, but it does not explain why. Remembering the equation $PV = nRT$ does not mean you have a good understanding of the gas laws.

3. **Scientists use models to try to better understand and explain concepts and ideas.** These models are not reality but a way of simplifying our ideas. What you need to do is to understand the models–do not merely memorize the premises of a model, but understand the significance and limitations of the model. Albert Einstein once said "Explanations should be as simple as possible, but not simpler." This is a good way of thinking about models–we want to maximize understanding and minimize complexity. Science is about making observations and using these to derive models that are further refined by new observations. The observations we make and the questions we want to answer guide our development of the model. For example, when discussing gas laws we generally think of gases as consisting of indestructible particles–no protons, electrons, or neutrons are in this model. Is this wrong? Yes and no. While atoms do consist of subatomic particles, these particles do not serve to increase our understanding of gas laws. Thinking of gases this way does not mean we are saying there are no subatomic particles, only that the model disregards them. A question that always comes up is, "If this model is not the

absolute truth, why do we bother studying it at all?" But this is what science is about. Unfortunately, we must develop simplified models because we simply cannot understand reality all at once. Models only become more complicated when we try to answer more questions. Our goal is to keep the model as simple as possible and answer as many questions as possible. When we find ourselves with too many questions we cannot answer, the model is refined and expanded or even disregarded. And it is acceptable that the model is not the "absolute truth" as long as you know this is the case and that you have an understanding of the limitations. Thus understanding the kinetic molecular theory, for example, goes beyond memorizing its premises; you should understand the premises, why they make sense, how they are simplified, what their limitations are, and their significance.

A question you are probably asking yourself at this point is, "What can I do to help myself do well in my chemistry course?" The quick answer is easy–you need to understand chemistry. But how do you do that?

To understand chemistry, you need to take an active role in your own learning. You have undoubtedly heard countless times from your parents and teachers that you are responsible for your own learning. But what does this mean in practice? You already go to class, do your homework, and study for quizzes and exams. What else can you do to take responsibility? One of the best ways to do this is to constantly ask questions. This includes while you are reading a text, attending a lecture, doing homework problems, reviewing your notes, or studying with friends. The key is that you consistently ask "Why?" and "What does this mean?" and "What are the implications?"

For example, when you are in lecture, listen intently to the proofs and derivations. Understand where the knowledge comes from and its implications and limits. Ask questions of the professor or at the very least write questions to yourself during lecture. You can think about these and ask the professor, a teaching assistant, or friends later.

When reading your text, do not read it as a novel. Be critical. Write down questions, look at the graphs and figures, and understand what they are telling you. You have heard the phrase "a picture is worth a thousand words," and this is true about graphs and figures. There is a lot of information in them, and you should extract and understand as much of it as possible. One goal of a text is to provide a source of detailed information and to slowly develop ideas. You are doing a disservice to yourself if you do not read it thoroughly and repeatedly, and you make learning chemistry more difficult.

In doing homework, use the *Solutions Manual* sparingly because the crucial part of the problems is setting them up. Many students view "doing problems" as getting them over with more than learning from them. Getting a correct answer does not necessarily imply understanding. Chemistry problems are not meant to be answered quickly and easily, but often require a lot of thought and trial and error. Think about what the problem is asking. Draw a picture. Do not just go directly to an equation, but think about what you have and where you are trying to go. Do not be afraid to make a mistake or to go down an incorrect pathway. Many times you will learn from this. Obviously you have a limited amount of time, and you cannot spend an inordinate amount of time on each problem. But students find that something interesting happens when they do problems this way. The problems seem to get easier, and the students find they can actually do fewer problems with more understanding.

LEARNING REVIEW

1. Explain why chemistry is important to you even if your career is far removed from the sciences.
2. Aside from helping you to get a good grade in chemistry, of what use are the problem-solving skills you will learn?

3. Imagine that you are a scientist exploring life on the newly discovered planet, Cryon. Cryon is cold and is perpetually covered with snow on one side. While exploring the snowy side of Cryon you repeatedly observe that all the birds have white feathers. You hypothesize that *all* the birds on Cryon have white feathers. Being a good scientist, you:
   a. Declare that all birds on Cryon must be white since all the ones on the snowy side are white.
   b. Test your hypothesis about all birds on Cryon being white by observing a bird color on the non-snow-covered part of the planet as well the snowy side.
   c. Elevate your hypothesis about white birds to a natural law that states that all life forms on cold planets that are covered with snow on one side are white.

ANSWERS TO LEARNING REVIEW

1. Chemistry will have a different impact on the career of each individual, but even if your career is far removed from the sciences, chemistry plays an important role in each of our everyday lives. We depend on the science of chemistry to provide us with a better standard of living.
2. Problem-solving skills can be used throughout your life. Many situations require you to think logically, to propose hypothetical solutions, and to choose the most reasonable one. Chemistry can help develop logical thinking skills.
3. You as the scientist in this problem have made some observations. But your information is not complete. You have no information about the color of birds on the other side of Cryon. A good scientist would test the hypothesis about bird color by collecting more data. It would not be appropriate to elevate the hypothesis to a natural law until the hypothesis was more thoroughly tested. Choice c is not correct; so the correct answer is b.

CHAPTER 2

Measurements and Calculations

INTRODUCTION

Chemistry is a science that requires observation of the world around us and measurements of the phenomena we observe. In this chapter you will learn how to record your observations and how to perform calculations with measured values. Scientific measurements are usually made using the metric system or the International System. You will need to become familiar with these systems of measurements and know the magnitude of each of the major units.

CHAPTER DISCUSSION

Significant Figures

Measurement is an important part of science, and an understanding of uncertainty is an important part of measurement. Science is often thought of (incorrectly) as a body of unchanging absolute truths, which makes the concept of uncertainty seem odd. But you should realize that uncertainty is always a factor in any measurement except for exact counting. For measurements you will be taking in the lab, there is always one (and only one) uncertain digit that we can reasonably estimate. Imagine, for example, measuring water in a beaker as shown below.

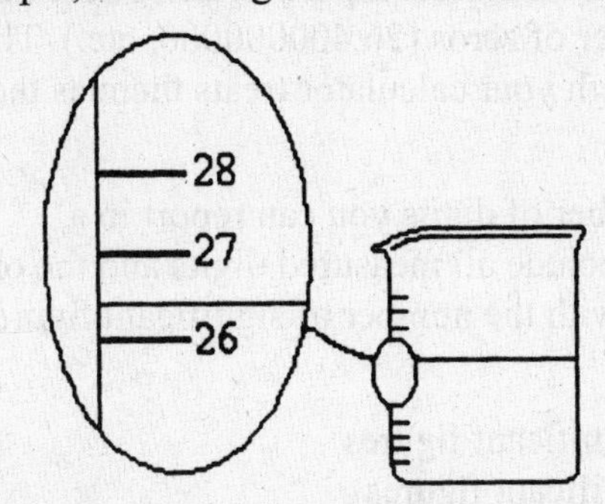

Using this beaker, we know there is more than 26 mL of water and less than 27 mL of water. To report "26 mL" or "27 mL" would be imprecise. Now imagine if we used a beaker as shown below.

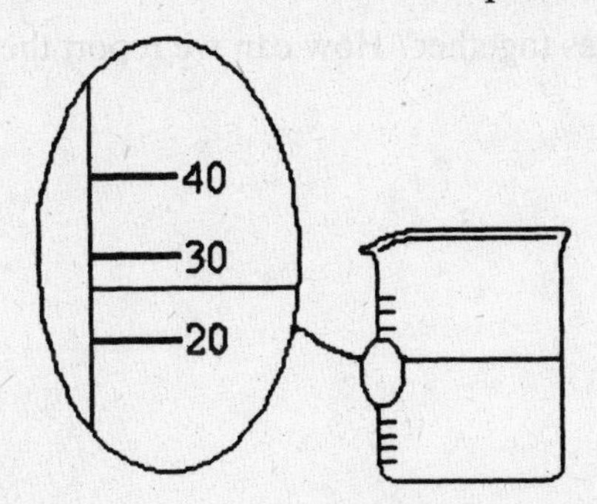

In this case we would report an answer to the ones place. In this case, the water appears just over the halfway point between "20" and "30," so "26" is a reasonable estimate. Note that we would not call this an exact measurement. The actual amount of water may be 25 mL or even 27 mL. Unless the glassware is marked, we generally assume our uncertainty is ±1 for the digit that we estimate.

Look back to the first beaker. We can make a reasonable estimate of the tenths place in this case. The water level appears to be just under halfway between the two graduations, so we might report 26.4 mL. In this case, we can assume that the actual amount of water is between 26.3 and 26.5 mL. Therefore we cannot report an answer of 26.42 mL since this would imply we knew the volume was between 26.41 and 26.43 mL (again, this assumes the glassware does not have a precision associated with it). What if we wanted to measure water to the hundredths place? This would require glassware with graduations as shown in the beaker below.

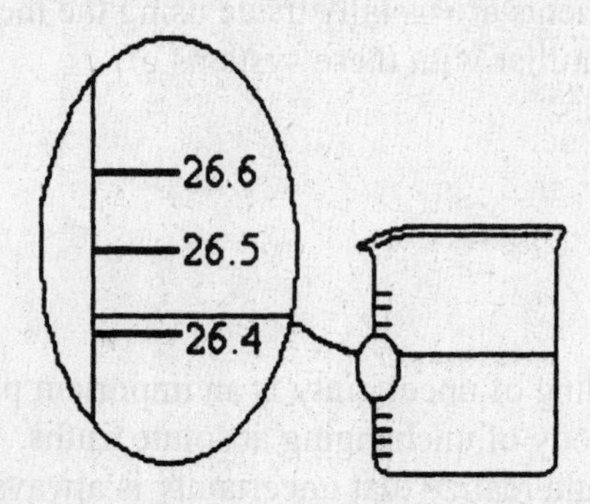

In this case, 26.42 mL is a reasonable estimate for the volume of water although you might think it is 26.41 or 26.43. Again, realize that we never get an exact measurement. Even if the water level seemed to be right on the 26.4 graduation, we would report 26.40 mL, but we cannot report "exactly 26.4 mL." Reporting "exactly 26.4 mL" implies 26.4 with an infinite number of zeros (26.400000000, etc.). Thus, 26.42 is not the same as 26.420 in terms of measurement (although your calculator treats them as the same). The only way to get an exact number is to count it.

Thus the glassware determines the precision that affects the number of digits you can report in a measurement. These digits are the significant figures, and they include all measured digits and the one estimated digit. Our three measurements in this example, along with the number of significant figures, are:

Beaker 1	26.2 mL	three significant figures
Beaker 2	26 mL	two significant figures
Beaker 3	26.42 mL	four significant figures

Now what happens if we add the water from each of these figures together? How can we report the results? Mathematically, we have:

$$\begin{array}{r} 26.4\ \text{mL} \\ 26\ \ \ \ \text{mL} \\ + \\ \underline{26.42\ \text{mL}} \\ 78.82\ \text{mL} \end{array}$$

However, we should realize that we have some uncertain digits. That is, the above procedure implies that the first measurement is 26.40, and the second is 26.00. However, this is simply not true. A better representation for this addition is

$$\begin{array}{r} 26.4?\ \text{mL} \\ 26.??\ \text{mL} \\ + \\ \underline{26.42\ \text{mL}} \\ 78.??\ \text{mL} \end{array}$$

Note in the hundredths we are adding a 2 to two unknown digits. What is "2 + ? + ?"? The answer has to be, "We don't know"! In this case, we know the sum only to the ones place, so we can only report it as such. So do we report it as 78? 79? Because the sum of the numbers is 78.82, we round up to 79. We can also justify this by recalling the uncertainty of the numbers. Let's assume two extreme cases. In the first case, assume we estimated too high for all three measurements (that is, assume there was actually less water than we thought). In the second case, assume we estimated too low for all three measurements. Remember that we assume we can be off by ±1 in the last digit. The range for the total amount of water in each case is shown below:

$$\begin{array}{r} 26.3\ \text{mL} \\ 25\ \ \ \ \text{mL} \\ +\ \underline{26.41\ \text{mL}} \\ 77.71\ \text{mL} \end{array} \qquad \begin{array}{r} 26.5\ \text{mL} \\ 27\ \ \ \ \text{mL} \\ +\ \underline{26.43\ \text{mL}} \\ 79.93\ \text{mL} \end{array}$$

The maximum range of volume should be between 77.71 mL and 79.93 mL. Since we can report the answer only to the ones place, the range should be between 78 mL and 80. mL. Therefore a reported answer of 79 mL (with a range of ±1) is reasonable.

After studying about measurement and significant figures, you should be able to answer the following questions:

1. Why do we care about significant figures? What is the point of determining which figures are significant? That is, what is the practical application?

2. Why is it that there is always one uncertain digit? Why can't we just measure more accurately? Why is there just one uncertain digit in the reported answer?

3. Make sense of the rules for which zeros are significant. Be able to explain them (not just recite) to a classmate or instructor. (One way to understand these is to relate the concept of significant figures to scientific notation).

Dimensional Analysis

When multiplying numbers in dimensional analysis, we are really just multiplying fractions. Remember, when multiplying fractions, multiply all of the numbers in the numerator first, followed by multiplying all of the numbers in the denominator. The last step is to divide the numerator product by the denominator product.

For example, to find the product of $\frac{1}{3}$ and $\frac{2}{5}$ we can write the expression in one of two ways:

$$\frac{1}{3} \times \frac{2}{5} \text{ or } \left(\frac{1}{3}\right)\left(\frac{2}{5}\right)$$

We then solve the problem using the following method:

$$\left(\frac{1}{3}\right)\left(\frac{2}{5}\right)=\frac{(1\times 2)}{(3\times 5)}=\frac{2}{15}=0.133$$

Whenever we see the same number in both the numerator and denominator, they cancel out (to equal 1).

$$\left(\frac{1}{\cancel{3}}\right)\left(\frac{\cancel{3}}{5}\right)=\frac{1}{5}=0.20$$

If the number we are analyzing is a whole number, remember that this really means that the number is over 1 (whole number in the numerator, 1 in the denominator). For example, the number 4 really means $\frac{4}{1}$.

When multiplying units, use the same principle that you use for multiplying fractions. If one unit is in the numerator, and the identical unit is in the denominator, they cancel each other out (and ultimately equal 1). Any remaining units are evaluated for the answer.

$$\left(\frac{\cancel{centimeter}}{1}\right)\left(\frac{meter}{\cancel{centimeter}}\right)=\frac{meter}{1}=meter$$

You can also multiply several units together at once using the same principle as for fractions.

$$\left(\frac{\cancel{centimeter}}{second}\right)\left(\frac{\cancel{meter}}{\cancel{centimeter}}\right)\left(\frac{\cancel{kilometer}}{\cancel{meter}}\right)\left(\frac{megameter}{\cancel{kilometer}}\right)=\frac{megameter}{second}$$

It is very important to note that if a unit appears once in the numerator but more than once in the denominator, we can cancel out only one of the unit expressions in the denominator. Think of this concept in terms of fractions. If there were the number 4 in the numerator and two 4's in the denominator of different fractions, we would cancel out only one of the 4's on the bottom, not both.

$$\left(\frac{\cancel{4}}{5}\right)\left(\frac{3}{\cancel{4}}\right)\left(\frac{1}{4}\right)=\frac{(3\times 1)}{(5\times 4)}=\frac{3}{20}=0.15$$

Let's look at an example with units. Consider multiplying the following units.

$$\left(\frac{kilogram}{second}\right)^2\left(\frac{meter}{kilogram}\right)^2\left(\frac{second}{meter}\right)=$$

The squared factor is equivalent to multiplying the fraction by itself.

$$\left(\frac{kilogram}{second}\right)\left(\frac{kilogram}{second}\right)\left(\frac{meter}{kilogram}\right)\left(\frac{meter}{kilogram}\right)\left(\frac{second}{meter}\right)=$$

Now we can evaluate the expression by canceling out units.

$$\left(\frac{\cancel{kilogram}}{second}\right)\left(\frac{\cancel{kilogram}}{\cancel{second}}\right)\left(\frac{meter}{\cancel{kilogram}}\right)\left(\frac{\cancel{meter}}{\cancel{kilogram}}\right)\left(\frac{\cancel{second}}{\cancel{meter}}\right)=\frac{meter}{second}$$

A Warning about Dimensional Analysis

Dimensional analysis is a double-edged sword. It is extremely useful and quite dangerous. It is dangerous because it can allow you to solve problems you do not understand. For example, consider the following problem:

> There are 2 igals in 1 odonku, and 6 odonkus in 4 falgers. If you have 3 igals, how many falgers is this?

We can solve this simply using dimensional analysis:

$$3 \text{ igals} \times \left(\frac{1 \text{ odonku}}{2 \text{ igals}}\right) \times \left(\frac{4 \text{ falgers}}{6 \text{ odonkus}}\right) = 1 \text{ falger}$$

Therefore, the answer is 1 falger. The questions to ask are "What is an igal? ," "What is an odonku?," "What is a falger? ," "What is the point of this problem?". Even though you can solve this problem, it is absolutely meaningless. And this is something you want to avoid in a chemistry course–solving problems without understanding them. Even if you can do this on some occasions, many of the problems in chemistry require an understanding of underlying principles, and it is good practice to start understanding early on. Dimensional analysis is a good tool for unit conversion, but you should never use it to try to replace understanding a problem.

LEARNING REVIEW

1. To express each of the following numbers in scientific notation, would you move the decimal point to the right or to the left? Would the power of 10 be positive or would it be negative (have a minus sign)?

   a. 0.001362

   b. 146,218

   c. 342.016

   d. 0.986

   e. 18.8

2. Complete the table below, and convert the numbers to scientific notation.

		Coefficient		Exponent
a.	0.00602	6.02	×	________
b.	60,000	6	×	________
c.	49	________	×	10^1
d.	1.002	1.002	×	________

3. Convert the numbers below to scientific notation.

   a. 1,999,945

   b. 650,700

   c. 0.1109

   d. 545

   e. 0.0068

   f. 0.042001

g. 1.2

h. 13.921

4. To express the following numbers in decimal notation, would you move the decimal point to the right or to the left? How many places?

a. 1.02×10^3

b. 4.1×10^{-6}

c. 5×10^5

d. 4.31×10^2

e. 9.31×10^{-2}

5. Convert the numbers below to decimal notation.

a. 4.91×10^{10}

b. 5.42×10^{-6}

c. 2.07×10^3

d. 1.009×10^{-4}

e. 9.2×10^1

f. 4.395×10^5

g. 7.03×10^{-2}

6. How can you convert −1235.1 to scientific notation?

7. Which quantity in each pair is larger?

a. 1 meter or 1 milliliter

b. 10 seconds or 1 microsecond

c. 1 centimeter or 1 millimeter

d. 1 kilogram or 1 decigram

8. Which quantity in each pair is larger?

a. 1 mile or 1 kilometer

b. 1 liter or 1 cubic meter

c. 1 kilogram or 1 pound

d. 1 quart or 1 milliliter

e. 1 micrometer or 12 inches

9. What metric or SI unit would you be likely to use in place of the English units given below?

a. Bathroom scales commonly provide weight in pounds.

b. A convenient way to purchase small quantities of milk is by the quart.

c. A cheesecake recipe calls for 1 teaspoon of vanilla extract.

d. Carpeting is usually priced by the square yard.

e. "An ounce of prevention is worth a pound of cure."

10. What number would you record for each of the following measurements?

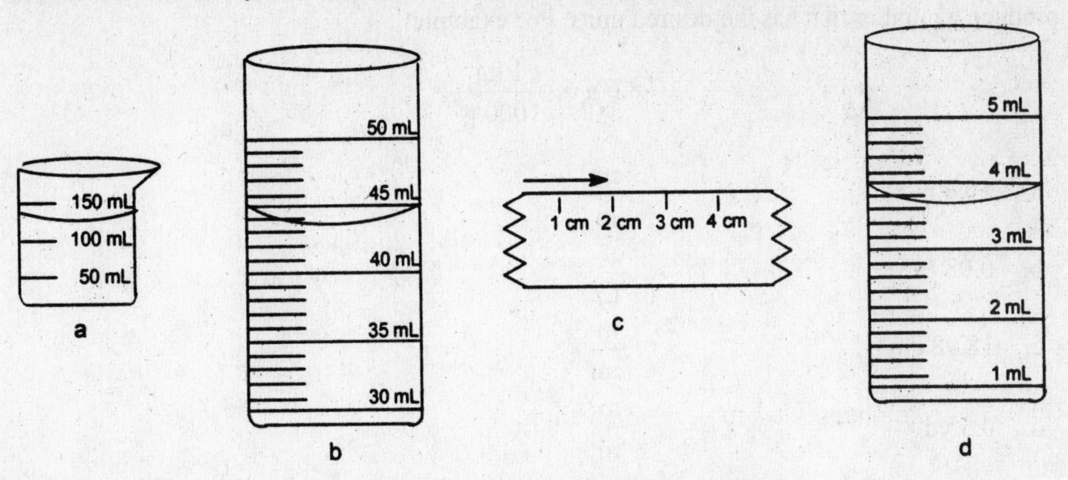

11. How many significant figures are in each of the following numbers?

    a. 100
    b. 1180.3
    c. 0.00198
    d. 1.001
    e. 67,342
    f. 0.0103
    g. 4.10×10^4

12. Express the results of each calculation to the correct number of significant figures.

    a. 1.8×2.93
    b. 0.002/0.041
    c. 0.00031×4.030
    d. 495.0/390
    e. 5024×19.2
    f. $91.3 \times 2.10 \times 7.7$
    g. $8.003 \times 4.93/61.05$

13. Round off the following numbers to the number of significant figures indicated.

	Number	Number of Significant Figures
a.	0.58333333	four
b.	451.0324	three
c.	942.359	four
d.	0.0090060	two
e.	6.8	one
f.	1346	three
g.	490,000.423	six
h.	0.06295	three

14. For each of the quantities below, give a conversion factor that will cancel the given units and produce a number that has the desired units. For example:

$$8.6\ \not{g} \times \frac{1\ \text{kg}}{1000\ \not{g}} =$$

a. $10.6\ \text{m} \times \frac{\text{cm}}{\text{m}}$

b. $0.98\ \text{L} \times \frac{\text{qt}}{\text{L}}$

c. $18.98\ \text{cm} \times \frac{\text{in}}{\text{cm}}$

d. $0.5\ \text{yd} \times \frac{\text{m}}{\text{yd}}$

e. $25.6\ \text{kg} \times \frac{\text{lb}}{\text{kg}}$

15. Perform the following conversions:

a. 5.43 kg to g

b. 65.5 in to cm

c. 0.62 L to ft^3

d. 111.3 g to lb

e. 40.0 qt to L

f. 2.83 g to lb

g. 0.21 cm to in

16. Fill in the important reference temperature on each of the temperature scales.

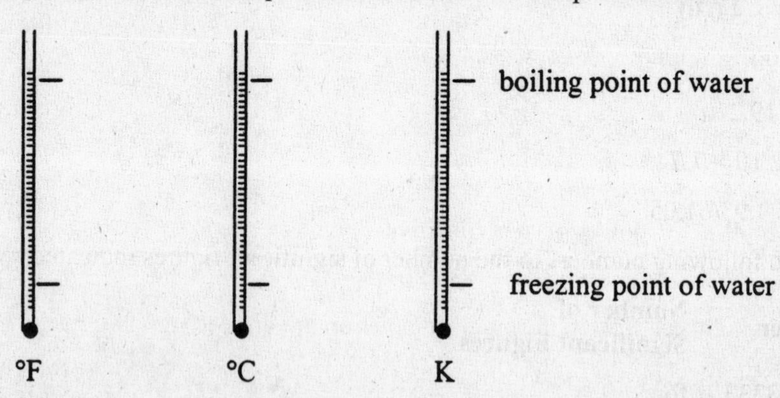

17. How many degrees are there between the freezing point and the boiling point of water on the Fahrenheit and on the Celsius scales?

Also:

a. Calculate the ratio of the number of degrees Fahrenheit to the number of degrees Celsius between the freezing and boiling points of water.

b. Calculate the ratio of the number of degrees Celsius to the number of degrees Kelvin between the freezing and boiling points of water.

c. Calculate the ratio of the number of degrees Fahrenheit to the number of degrees Kelvin between the freezing and boiling points of water.

18. Comfortable room temperature for houses is 75 °F. What is this on the Celsius scale?

19. Ethyl alcohol boils at 78.0 °C. What is this on the Fahrenheit scale?

20. In some parts of the Midwest, temperatures may drop as low as −22 °F in winter. What is this on the Kelvin scale?

21. Perform the temperature conversions below.

a. 180 °F to °C

b. −10.8 °C to K

c. 244 K to °C

d. 25.1 °F to °C

22. Fill in the missing quantities in the table below.

Substance	Density (g/mL)	Mass	Volume
seawater	1.025	52.6 g	________
diamond	________	2.13 g	0.65 mL
beeswax	0.96	125.5 g	________
oak wood	________	4.63 g	6173.3 mL

ANSWERS TO LEARNING REVIEW

1. To convert to scientific notation for numbers that are greater than zero but less than one, move the decimal point to the *right*. For numbers that are greater than one, move the decimal point to the *left*. Make sure your final answer has only one number to the left of the decimal point.

a. right 0.001362 power of ten: negative

b. left 146218 power of ten: positive

c. left 342.016 power of ten: positive

d. right 0.986 power of ten: negative

e. left 18.8 power of ten: positive

2. Remember that numbers written in scientific notation are divided into two parts. The coefficient on the left is a small number between one and ten, and the exponent on the right is ten raised to some power.

	Coefficient		Exponent
a. 0.00602	6.02	×	10^{-3}
b. 60,000	6	×	10^{4}
c. 49	4.9	×	10^{1}
d. 1.002	1.002	×	10^{0}

3. The answer for g, 1.2×10^0, means that we do not need to move the decimal point of the coefficient. 1.2×10^0 is the same as writing 1.2.

   a. 1.999945×10^6

   b. 6.507×10^5

   c. 1.109×10^{-1}

   d. 5.45×10^2

   e. 6.8×10^{-3}

   f. 4.2001×10^{-2}

   g. 1.2×10^0

   h. 1.3921×10^1

4. When converting from scientific notation to decimal, look first at the exponent. If the exponent is positive (has no negative sign), move the decimal point to the right. If the exponent is negative, move the decimal point to the left.

   a. right 1020

   b. left 0.0000041

   c. right 500,000

   d. right 431

   e. left 0.0931

5. A large number such as 49,100,000,000 has only three significant figures. The trailing zeros are not significant because there is no decimal point at the end.

   a. 49,100,000,000

   b. 0.00000542

   c. 2070

   d. 0.0001009

   e. 92

   f. 439,500

   g. 0.0703

6. This number is different from others we have seen. It is smaller than one and also smaller than zero. You can convert these numbers to scientific notation in much the same way as you convert numbers that are greater than one. First move the decimal point to the left as you normally would.

   $$-1235.1$$

   Then count the number of times the decimal point was moved, and add the correct exponent.

   $$1.2351 \times 10^3$$

   Just keep the minus sign in front of the entire number.

   $$-1.2351 \times 10^3$$

The minus sign goes in front of 1.235 because this number is less than zero. The exponent is negative only for numbers that are between zero and one.

7. To work this problem you need to have learned the SI prefixes and how they modify the size of the base unit.

   a. A meter is larger than a millimeter.

   b. 10 seconds is larger than 1 microsecond.

   c. 1 Mm is larger than 1 cm.

   d. 1 kilogram is larger than 1 decigram.

8. This problem asks about the relationship between English units and SI units. You need to know the relative sizes of English and SI units.

   a. 1 mile is larger than 1 kilometer.

   b. 1 cubic meter is larger than 1 liter.

   c. 1 kilogram is larger than 1 pound.

   d. 1 quart is larger than 1 milliliter.

   e. 12 inches is larger than 1 micrometer.

9.

   a. kilograms

   b. liter

   c. milliliter

   d. square meter (m^2)

   e. "A gram of prevention is worth a kilogram of cure."

10.

   a. This measuring device is a beaker. Each division represents 50 mL. The volume of liquid in the beaker is somewhere between 100 mL and 150 mL. We estimate that the volume is 120 mL.

   b. This measuring device is a graduated cylinder. The numbers tell us that each major graduation is 5 mL, so each of the smaller lines must be 1 mL. We can accurately measure the volume to the nearest 1 mL. The volume in this cylinder is between 43 and 44 mL. We estimate the volume to be 43.5 mL.

   c. The length of the arrow lies between 1 cm and 2 cm. We estimate that the arrow lies 0.9 of the way between the two marks. So the reported measurement would be 1.9 cm.

   d. This graduated cylinder has major divisions of 1 mL. The smaller marks represent 0.2 mL. The liquid lies between 3.6 and 3.8 mL. We estimate that the volume is about a quarter (0.05) of the way between the two marks, so the volume would be reported as 3.65 mL.

11. Remember that all nonzero numbers count as significant figures, and zeros in the middle of a number are always significant. Zeros to the right of some nonzero numbers are significant only if they are followed by a decimal point.

   a. 1

   b. 5

c. 3

d. 4

e. 5

f. 3

g. 3

12. For problems involving multiplication and division, your answer should have the same number of decimal points as the measurement with the least number of significant figures. For problems involving addition and subtraction, your answer should have the same number of significant figures as the measurement with the least number of digits to the right of the decimal point.

a. 5.3

b. 0.05

c. 0.0012

d. 1.3

e. 96,500

f. 1500

g. 0.646

13. You can answer problems such as 13.g by putting the decimal point at the end to show that all six digits are significant or use scientific notation with a coefficient that contains six digits.

a. 0.5833

b. 451

c. 942.4

d. 0.0090

e. 7

f. 1350

g. 490,000. or 4.90000×10^5

h. 0.0630

14. To answer this question, you need to know the common equivalencies and how to write them as a unit factor.

a. $10.6 \text{ m} \times \dfrac{100 \text{ cm}}{1 \text{ m}}$

b. $0.98 \text{ L} \times \dfrac{1.06 \text{ qt}}{1 \text{ L}}$

c. $18.98 \text{ cm} \times \dfrac{1 \text{ in}}{2.54 \text{ cm}}$

d. $0.5 \text{ yd} \times \dfrac{1 \text{ m}}{1.094 \text{ yd}}$

e. $25.6 \text{ kg} \times \dfrac{1000 \text{ g}}{1 \text{ Kg}}$

15.

a. $5.43 \text{ kg} \times \dfrac{1000 \text{ g}}{1 \text{ kg}} = 5430 \text{ g}$

b. $65.5 \text{ in} \times \dfrac{2.54 \text{ cm}}{1 \text{ in}} = 166 \text{ cm}$

c. $0.62 \text{ L} \times \dfrac{1 \text{ ft}^3}{28.32 \text{ L}} = 0.022 \text{ ft}^3$

d. $111.3 \text{ g} \times \dfrac{1 \text{ lb}}{453.6 \text{ g}} = 0.2454 \text{ lb}$

e. $40.0 \text{ qt} \times \dfrac{1 \text{ L}}{1.06 \text{ qt}} = 38 \text{ L}$

f. $2.83 \text{ g} \times \dfrac{1 \text{ lb}}{453.6 \text{ g}} = 6.24 \times 10^{-3} \text{ lb}$

g. $0.21 \text{ cm} \times \dfrac{1 \text{ in}}{2.54 \text{ cm}} = 0.083 \text{ in}$

16.

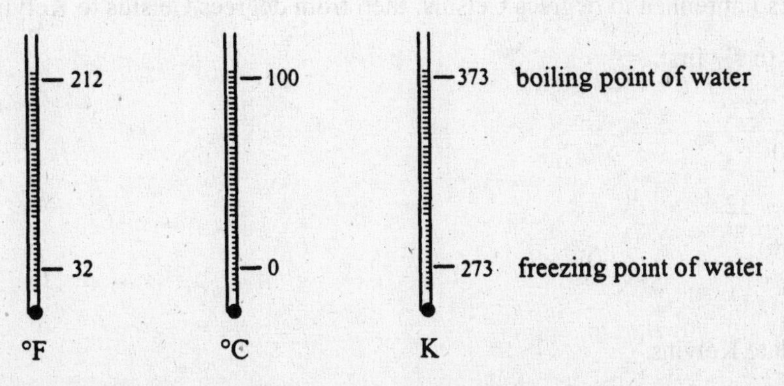

17. There are 180 degrees between the freezing and boiling points of water on the Fahrenheit scale and 100 degrees on the Celsius scale.

a. $\dfrac{°F}{°C} = \dfrac{180}{100} = 1.80$

b. $\dfrac{°C}{K} = \dfrac{100}{100} = 1$

c. $\dfrac{°F}{K} = \dfrac{180}{100} = 1.80$

18. We want to convert from degrees Fahrenheit to degrees Celsius.

$T_{°F} = 75$

We can use the formula below to calculate degrees Celsius.

$$T_{°C} = \frac{T_{°F} - 32}{1.80} = \frac{75 - 32}{1.80}$$

$T_{°C} = 24$

75 degrees Fahrenheit is equivalent to 24 degrees Celsius.

19. We want to convert from degrees Celsius to degrees Fahrenheit.

$T_{°C} = 78.0$

We can use the formula below to calculate degrees Fahrenheit.

$T_{°F} = 1.80\,(T_{°C}) + 32$

$T_{°F} = 1.80(78.0) + 32$

$T_{°F} = 172$

78.0 degrees Celsius is equivalent to 172 degrees Fahrenheit.

20. We want to convert from degrees Fahrenheit to Kelvin.

$T_{°F} = -22$

We do not have a formula to directly convert degrees Fahrenheit to Kelvins, but we can convert from degrees Fahrenheit to degrees Celsius, then from degrees Celsius to Kelvins.

Convert $T_{°F}$ to $T_{°C}$ first.

$$T_{°C} = \frac{T_{°F} - 32}{1.80}$$

$$T_{°C} = \frac{-22 - 32}{1.80}$$

$T_{°C} = -30.$

Now, calculate Kelvins.

$T_K = T_{°C} + 273$

$T_K = -30. + 273$

$T_K = 243$

21.

a. $T_{°F} = 180$

$$T_{°C} = \frac{T_{°F} - 32}{1.80}$$

$$T_{°C} = \frac{180 - 32}{1.80}$$

$T_{°C} = 82$

b. $T_{°C} = -10.8$

$T_K = T_{°C} + 273$

$T_K = -10.8 + 273$

$T_K = 262$

c. $T_K = 244$

$T_K = T_{°C} + 273$

Rearrange this equation to isolate $T_{°C}$.

$T_{°C} = T_K - 273$

$T_{°C} = 244 - 273$

$T_{°C} = -29$

d. $T_{°F} = 25.1$

$$T_{°C} = \frac{T_{°F} - 32}{1.80}$$

$$T_{°C} = \frac{25.1 - 32}{1.80}$$

$T_{°C} = -3.8$

22.

Substance	Density	Mass	Volume
seawater	1.025 g/mL	52.6 g	51.3 mL
diamond	3.3 g/mL	2.13 g	0.65 mL
beeswax	0.96 g/mL	125.5 g	130 mL
oak wood	0.750 g/mL	4.63 g	6173.3 mL

CHAPTER 3

Matter

INTRODUCTION

This chapter provides you with a basic foundation of facts and concepts about matter you will need throughout your chemistry course. There are fewer mathematical calculations in this chapter than in other chapters you will study.

Pay careful attention to Section 3.2 (Physical and Chemical Properties and Changes) and 3.4 (Mixtures and Pure Substances) in your textbook. The concepts in these sections often seem confusing when you are first introduced to them. Look carefully at the examples in your text that will help you distinguish between physical and chemical changes and between mixtures and pure substances.

CHAPTER DISCUSSION

Take the opportunity while studying this chapter to get used to thinking microscopically; that is, at a molecular level. One of the most difficult aspects of learning chemistry is that we see on a large scale (a macroscopic level) but the chemical changes and physical processes occur at a molecular level (microscopic level). You are expected to be able to relate the two of these. See Figure 3.2 in your text for the difference between these two perspectives. Atoms and molecules will be formally introduced in Chapter 4, but it is a good idea to start thinking microscopically now.

For example, understanding the difference between physical changes and chemical changes requires thinking at a molecular level. In Section 3.2 of your text, the phrase "change in composition" is used to denote a chemical change. But what does this mean? Steam appears to us to be vastly different from ice. Do they have different compositions? To understand this, we need to know what the term "composition" means. For a chemist, composition of a substance has to do with the makeup of the molecules. Thus, heating ice until it melts and heating the water until it boils does not change the molecules. The substance is still made of water molecules, each with two atoms of hydrogen and one atom of oxygen and symbolized H_2O. Note that in Figure 3.2 ice, liquid water, and steam are all made of H_2O molecules. Thus melting, freezing, boiling, and condensing are all physical processes–the molecules are left unchanged.

If water underwent a chemical change, however, its composition (molecular makeup) would change. The bonds holding the hydrogen and the oxygen atoms together in a water molecule would break, and new bonds would form, making hydrogen gas (H_2) and oxygen gas (O_2). We can visualize this chemical change with the following representation:

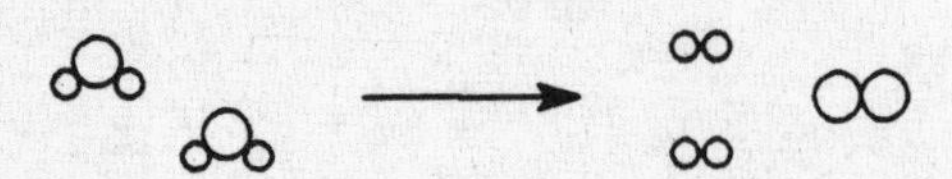

Water (H_2O) has a different composition from a mixture of hydrogen and oxygen (H_2 and O_2) and different chemical properties. For example, putting a lit match to a mixture of hydrogen and oxygen can result in a loud explosion (do not try this), but putting a lit match to water results in a wet match.

Therefore, making sense of chemical and physical changes is difficult without a molecular perspective. The same is true with the concepts of elements, mixtures, and compounds. Don't merely memorize the text definitions for these, but instead understand the differences among each of these (including heterogeneous and homogeneous mixtures). One of the best ways to achieve this understanding is to make sketches of these at a molecular level.

For example, answer the following questions (think about these before reading on).

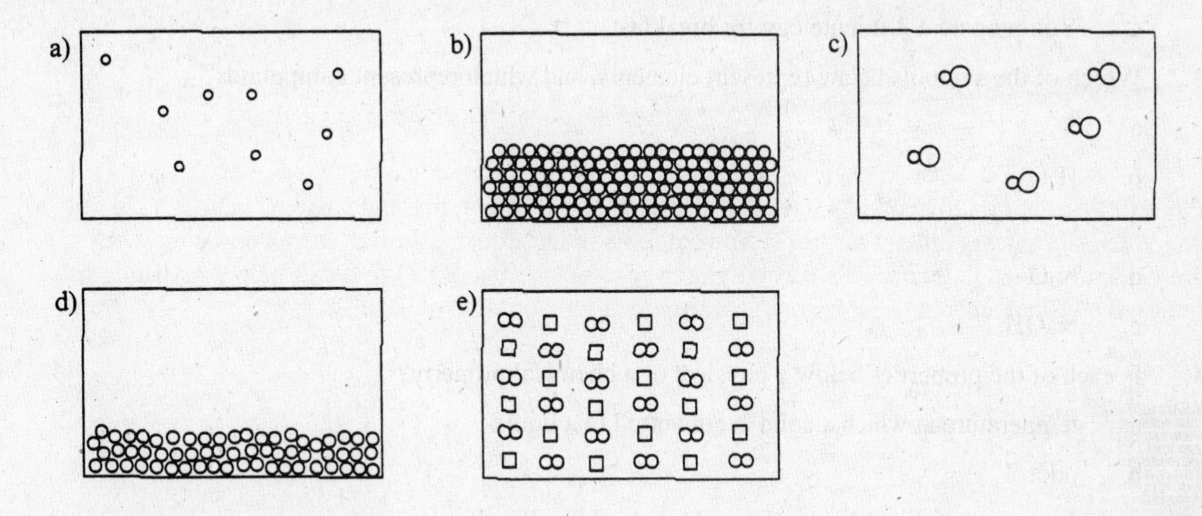

1. Which best represents a homogeneous mixture of an element and a compound?
2. Which best represents a gaseous compound?
3. Which best represents a solid element?
4. Which best represents a heterogeneous mixture of two elements?
5. What would you term the choice not chosen in 1-4?

The answers are "e," "c," "b," "d" and "a gaseous element." Notice that a compound differs from an element in that a compound is made of at least two different types of atoms.

You should be able to draw molecular-level sketches of any possible combinations of elements, compounds, mixtures (homogeneous and heterogeneous), gases, liquids, and solids. For example, sketch a homogeneous mixture of a gaseous element in a liquid compound.

You also should be able to answer the following questions:

1. What is wrong with the term "heterogeneous compound"? "Homogeneous compound"?
2. Sketch molecular-level diagrams to make sense of Figure 3.9 and Figure 3.10.

LEARNING REVIEW

1. Which of the properties below is/are physical properties, and which is/are chemical properties?
   a. Oxygen atoms can combine with hydrogen atoms to form water molecules.
   b. Ethyl alcohol boils at 78°C.
   c. Liquid oxygen is pale blue in color.

2. Which of the changes below are physical changes, and which are chemical changes?
   a. A copper strip is hammered flat to make a bracelet.
   b. Copper and sulfur react to form a new substance, copper(I) sulfide.
   c. Liquid water freezes at 273 K.
   d. Oxygen gas condenses to a liquid at −183°C.
   e. You prepare a 3-minute egg for breakfast.
3. Which of the symbols below represent elements, and which represent compounds?
   a. S
   b. H_2O
   c. C
   d. N_2O_5
   e. NaOH
4. Is each of the properties below a physical or a chemical property?
   a. temperature at which a solid is converted to a liquid
   b. odor
   c. temperature at which a compound breaks down into its elements
   d. oxygen reacts with a substance to produce energy
5. Which of the substances below are mixtures, and which are pure substances?
   a. gasoline
   b. table sugar (sucrose)
   c. garden soil
   d. sterling-silver necklace
6. Which of the mixtures is homogeneous and which is heterogeneous?
   a. sweetened hot tea
   b. plastic bag filled with leaves and grass clippings
   c. a weak solution of rubbing alcohol in water
   d. devil's food cake mix
7. Describe how you can separate a mixture by filtration.
8. What type of mixture is best separated by filtration, a homogeneous mixture or a heterogeneous mixture?
9. Describe how you would separate the following mixtures.
   a. sand from gravel
   b. salt from sand
   c. sugar from water

10. Label each part of the distillation apparatus below.

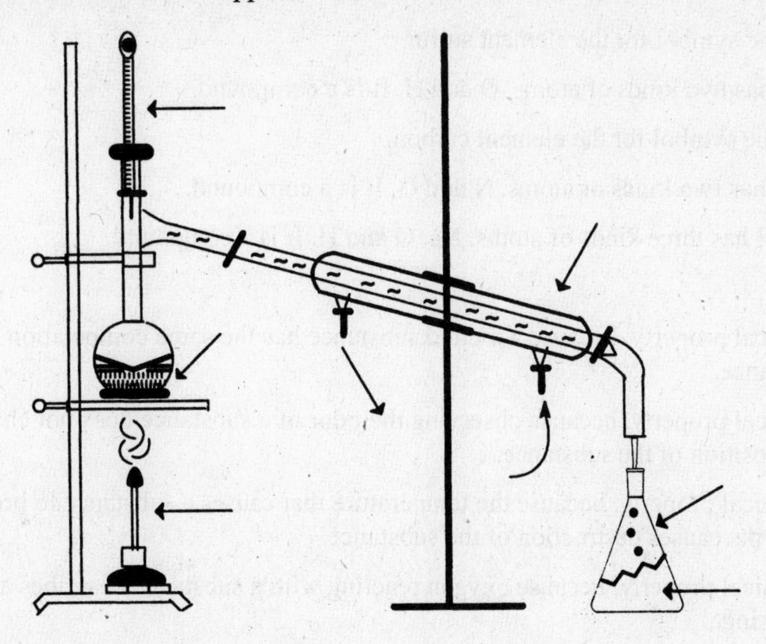

ANSWERS TO LEARNING REVIEW

1.

a. Oxygen can combine with hydrogen to produce a new substance, water. Because oxygen and hydrogen have the potential to combine to form a new substance, this is an example of a chemical property.

b. Observing ethyl alcohol boiling does not destroy or change the ethyl alcohol molecules. Therefore boiling point is a physical property.

c. Observing the color of a substance does not change its composition. The pale blue color of liquid oxygen is a physical property.

2.

a. When a copper strip is hammered into a bracelet, the shape of the copper is changed, but not the composition. This is a physical change.

b. The new substance, copper sulfide, is a black solid that does not have any of the characteristics of copper metal or yellow elemental sulfur. This is a chemical change.

c. When liquid water freezes to become solid ice, the molecules are still those of water. No change in composition has occurred. This is a physical change.

d. Liquid oxygen molecules become solid oxygen molecules at −183°C. This is a physical change.

e. Cooking an egg causes changes to the egg's composition. Heating changes the structure of large egg proteins, causing them to form solids. This is a chemical change.

3.

a. S is the symbol for the element sulfur.

b. H_2O has two kinds of atoms, O and H. It is a compound.

c. C is the symbol for the element carbon.

d. N_2O_5 has two kinds of atoms, N and O. It is a compound.

e. NaOH has three kinds of atoms, Na, O and H. It is a compound.

4.

a. Physical property, because a melted substance has the same composition as a solid substance.

b. Physical property, because observing the odor of a substance does not change the composition of the substance.

c. Chemical property, because the temperature that causes a substance to break down into elements causes destruction of the substance.

d. Chemical property, because oxygen reacting with a substance describes a chemical change occurring.

5.

a. Gasoline is a mixture of complex carbon-containing molecules, detergents, and additives.

b. Table sugar is a pure substance. It contains only sucrose molecules.

c. Garden soil is a mixture. It contains sand, water, clay, dead plant leaves, and other components.

d. Sterling silver is made from 93% silver and 7% copper by mass. It is a mixture.

6.

a. Sweetened tea is a homogeneous mixture. When sugar is added to hot tea, it dissolves. All of the tea is equally sweetened.

b. A bag full of leaves and grass clippings is a heterogeneous mixture. Some parts of the bag will contain more leaves than grass, and other parts will contain more grass than leaves.

c. Rubbing alcohol and water mix freely with each other. The molecules of one completely disperse in molecules of the other. This produces a homogeneous solution.

d. A devil's food cake mix is a homogeneous mixture. There are no lumps (usually) of sugar, or clumps of flour. All the ingredients are distributed equally throughout the mix.

7. Separating mixtures by filtration depends upon differences in the physical properties of the components of the mixture. Pour the mixture onto a mesh. One common mesh is filter paper, which is made from a mesh of cellulose fibers. The liquid in the mixture can pass through the cellulose fibers into a container below. The particles remain behind, trapped in the fibers of the mesh.

8. Filtration can separate a heterogeneous mixture that contains a liquid and a solid component.

9.

a. Sand and gravel can be separated by filtration. Use a mesh that allows the sand particles to pass through but retains the gravel, for example, a piece of wire screen.

b. A salt and sand mixture can be separated by adding water and filtering. The salt will dissolve in the water, while the sand will not. By filtering the sand and salt water mixture, the pure sand remains behind while the salt passes through the mesh with the water.

c. Sugar and water cannot be separated by filtration because the sugar molecules dissolve in the water and will pass through a mesh. They can, however, be separated by distillation. Heat the mixture until the water boils (at 100 °C) and is converted to steam. The sugar has a higher boiling point than water does and is not vaporized. The steam rises in the distillation apparatus and can be captured and recondensed to liquid water. The sugar remains behind in the distillation flask.

10.

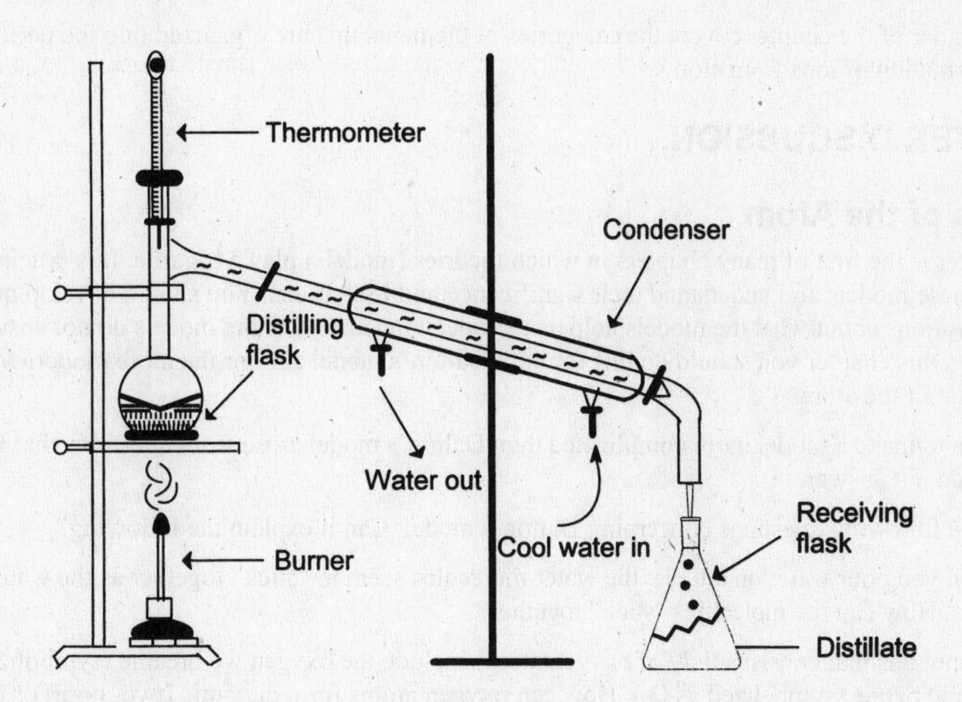

CHAPTER 4

Chemical Foundations: Elements, Atoms, and Ions

INTRODUCTION

In Chapter 4 you are introduced to the names and symbols of the common elements and ions. Make sure you learn the names and symbols now. Most of the chemistry covered in subsequent chapters depends upon knowledge of these names and symbols.

The remainder of the chapter covers the categories of elements that are organized into the periodic table and the formation of ions from atoms.

CHAPTER DISCUSSION

Models of the Atom

This chapter is the first of many chapters in which theories (models) play a big role. It is crucial to examine these models and understand their significance and limitations. You should develop questions and observations about what the models help us to understand and what the models do not answer. For example, in this chapter you should do this for both Dalton's model and for the more modern view of the structure of the atom.

The reason to make a model more complicated than Dalton's model arose from questions that Dalton's model could not answer.

Look at the following questions concerning Dalton's model. Can it explain the following?

1. When you pour water on a table, the water molecules seem to "stick" together as the water forms drops. How can the molecules "stick" together?

2. Compounds that consist solely of oxygen atoms include the oxygen we breathe (symbolized as O_2) and ozone (symbolized as O_3). How can oxygen atoms form diatomic (two-atom) or triatomic (three-atom) molecules? Why are other molecules with only oxygen unstable?

3. Chlorine naturally exists as a yellow-green gas. Nitrogen naturally exists as a colorless gas. Iron naturally exits as a solid. How can these different properties be explained?

Realize that Dalton's simple model cannot account for the observations made above. We know, then, that Dalton's model is incomplete. However, Dalton's model is still useful (for example, when we study gas laws in Chapter 13, we will use it almost exclusively). It is important to understand not only what the model tells us, but what it doesn't answer.

The next models of the atoms began incorporating the subatomic particles protons, neutrons, and electrons. But even this does not answer all of our questions. For example, in Section 4.6 you will find the question, "If all atoms are composed of these same components, why do different atoms have different chemical properties?" This requires a more modern view of the structure of the atom.

Even this modern view of the atom as presented in this chapter (and expanded upon in Chapter 11) brings about questions. Consider the following:

> Opposite charges are said to attract one another, and like charges are said to repel. If this is the case, why do the electrons not fall into the nucleus? What holds the protons together in the nucleus?

The simple models that are discussed in Chapter 4, while more complicated than Dalton's model (and while they can answer more questions than Dalton's model) cannot answer these questions.

And even though Dalton's model is quite simple and brings about many questions, we still use it frequently. Look back, for example, at the molecular level drawings in Chapter 3 (and these drawings throughout the text). They use Dalton's model because it conveys all the useful information. For example, consider chemical formulas in the next section. This emphasizes once again that all models are simplifications and will fail at some point. On the other hand, models are also extremely useful when properly applied.

Chemical Formulas

A chemical formula symbolizes the makeup of a molecule. For example, we can sketch a molecule of water as

However, it is much easier to write "H_2O," which conveys that there are two hydrogen atoms joined with one oxygen atom in a water molecule. Note, though, that the formula "H_2O" does not tell us that the order of atoms is HOH. In most cases, the formula just tells you which atoms, and how many of each, are in the molecule. So how do we know the structure? Sometimes formulas will be written to indicate the structure. Also, we will study how to determine the structure from a formula in more detail in Chapter 12.

To test your understanding of this idea, which of the following sketches do you think best represents "$2NH_3$"? Think about this before reading on.

a) 

b) 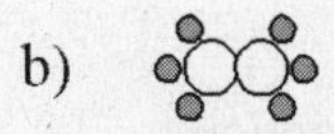

c)

The correct answer is "c." The ammonia molecule (NH_3) has one nitrogen atom and three hydrogen atoms bonded together. The "2" in front of the "NH_3" just means we have two of these molecules. We will see this idea again in Chapter 6–the "2" is called a coefficient, and the "3" is a subscript. We can examine this further by looking at Figure 4.15 in your text. Notice how a chemical equation (which we will study in more detail in Chapter 6) can be symbolized with a molecular perspective (that, incidentally, uses Dalton's model). You can symbolize the chemical reaction shown in Figure 4.15 as

$$2H_2O \rightarrow 2H_2 + O_2$$

Make sense of these symbols and how they relate to the molecular-level representation, and you are in good shape for understanding Chapter 6 (which is crucial for understanding the remaining chapters, especially Chapters 7, 8 and 9).

The Periodic Table

This chapter also introduces you to the periodic table that will be studied in more detail in Chapter 11. For now you should realize that the periodic table was constructed to minimize confusion and memorization and increase understanding and explanation, so use it this way. The elements in the table are not arranged alphabetically or chronologically according to discovery, but according to the number of protons. The elements in the same vertical column have chemical similarities that you will study later. But you should realize now that you can determine the most stable charge of many of the ions made from atoms.

For example, look at the example problems in Section 4.11 of your text. Notice that the most stable ionic charge for Na and Li is 1+, and both are in the same column of the periodic table. What is the most stable charge for Mg and Ca as ions? Each stable ion has a charge of 2+, each of these is in the same column, and they are one column from Li and Na. Find other examples of this, and realize that the periodic table is loaded with such patterns. Now is the time to begin your appreciation of this wonderful table, which provides many answers. Use the periodic table as the valuable resource that it is.

LEARNING REVIEW

1. This review question can help you determine your progress with the material in Chapter 4. You should be able to answer each of the questions below for the common elements listed in Table 4.3 of the textbook. Answer each question below for the element symbolized by Br.

   a. What is the name of the element?

   b. In which group of the periodic table is it found?

   c. What is its family name?

   d. When found in nature uncombined with other elements, what is its state?

   e. At room temperature, what is its physical state; solid, liquid, or gas?

   f. What is the name and charge of the ion it forms?

   g. How many neutrons are found in this isotope $^{80}_{35}Br$?

2. Which of the ten most abundant elements (determined by mass percent) on earth are not found in large amounts in the human body?

3. Match the elements below with the correct description.

oxygen	most abundant element on earth
silicon	most abundant element in the human body
carbon	trace element in human body
titanium	25.7% of mass on earth
hydrogen	these three elements make up 93% of mass in the human body
molybdenum	less than 1% of the mass on earth

4. Write symbols for the following elements.

   a. arsenic

   b. fluorine

   c. magnesium

   d. iron

e. neon

f. lead

g. potassium

h. chromium

i. nitrogen

j. calcium

5. Which of the common elements in Table 4.3 of your textbook have a one-letter symbol?

6. Some of the element symbols are not related to the modern name of the element. What are the elements represented by the following element symbols?

a. W

b. Hg

c. Cu

d. K

e. Fe

f. Pb

g. Sb

h. Na

7. Match the element name with the correct element symbol.

cadmium	Cl
carbon	Cr
calcium	C
chlorine	Co
cobalt	Cu
copper	Cd
chromium	Ca

8. Match the element symbol with the correct element name.

Na	silver
Sr	sulfur
S	sodium
Ag	silicon
Si	strontium

9. Describe the main parts of Dalton's atomic theory.

10. How does Dalton's atomic theory relate to the law of constant composition?

11. Dalton's model became more widely accepted when the existence of NO, NO_2, and N_2O became known. What aspect of Dalton's model allowed Dalton to predict the existence of these compounds?

12. Write chemical formulas for the following compounds.
    a. ethyl alcohol, which contains two carbon atoms, six hydrogen atoms, and one oxygen atom
    b. a compound that contains one atom of magnesium and two atoms of bromine
    c. a compound that contains four atoms of phosphorus and ten atoms of oxygen
    d. a compound that contains one atom of arsenic and three atoms of hydrogen

13. What is the *total* number of atoms found in each of the following compounds? What is the total number of elements found in each?
    a. KOH
    b. N_2O_3
    c. CCl_4
    d. H_2O_2
    e. Na_3PO_4

14. A physicist named J. J. Thomson showed that all atoms can be made to emit tiny particles that are repelled by the negative pole of an electric field. Which subatomic particle was this evidence for?
    a. proton
    b. neutron
    c. electron
    d. nucleus
    e. isotope

15. Match the scientist with the discovery.

Ernest Rutherford	demonstrated the existence of electrons
J. J. Thomson	demonstrated the existence of neutrons
Lord Kelvin	developed the plum pudding model of the atom
Rutherford & Chadwick	developed the nuclear-atom model from gold-foil experiments

16. Label the parts of the experimental apparatus used to develop the model of the nuclear atom.

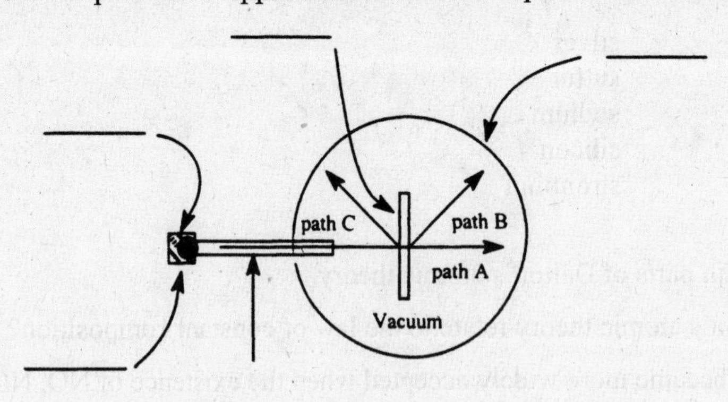

17. In the gold-foil experiment, how did Rutherford interpret each of the following observations?

    a. Most of the α-particles traveled unimpeded through the foil.

    b. Some of the α-particles were deflected slightly from the straight path when they entered the foil.

    c. A few of the α-particles bounced back when they entered the foil.

18. Fill in the missing relative masses and relative charges for each of the subatomic particles.

	Relative Mass	Relative Charge
a. Electron	________	1 –
b. Proton	________	________
c. Neutron	1839	________

19. Is the following statement true or false? An isotope of sodium could contain 12 protons, 12 neutrons and 11 electrons.

20. Label the parts of the symbol (X) below.

    _____ *A*

    X _____

    _____ *Z*

21. Write the symbols for the isotopes below in ${}^{A}_{Z}X$ notation.

    a. An isotope of hydrogen has an atomic number of 1 and a mass number of 3.

    b. An isotope of chlorine has an atomic number of 17 and a mass number of 37.

    c. An isotope of oxygen has 8 protons and 10 neutrons.

    d. An isotope of uranium has 92 electrons and 143 neutrons.

    e. An isotope of sulfur has an atomic number of 16 and 16 neutrons.

22. An isotope of titanium contains 24 neutrons and has a mass number of 46.

    a. How many protons does it contain?

    b. How many electrons does it contain?

23. Aluminum-29 has an atomic number of 13.

    a. What is its mass number?

    b. How many neutrons does it have?

24. Match the group name on the left with an element found in that group.

halogen	Ca
transition metal	Ne
alkali metal	Fe
alkaline earth metal	K
noble gas	F

25. Fill in the boxes of the periodic table with element symbols for each of the families below. The number at the top of each box represents atomic number.

a. **halogens**

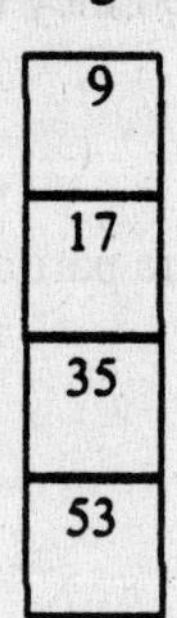

b. **alkaline earth metals**

c. **noble gases**

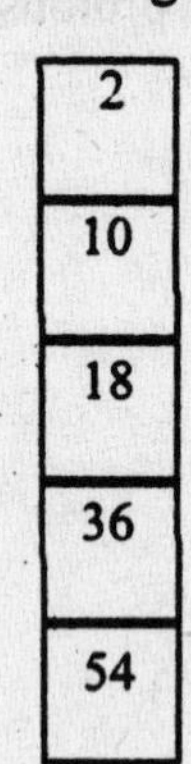

d. **alkali metals**

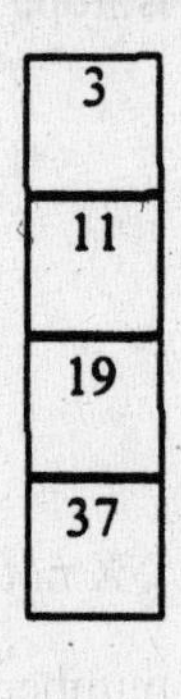

26. Which of the following elements are nonmetals?

a. Al

b. C

c. Cr

d. P

e. Br

f. I

27. Some of the elements along the jagged line on the right side of the periodic table have properties of both metals and nonmetals. Fill in the elemental symbols for these metalloids.

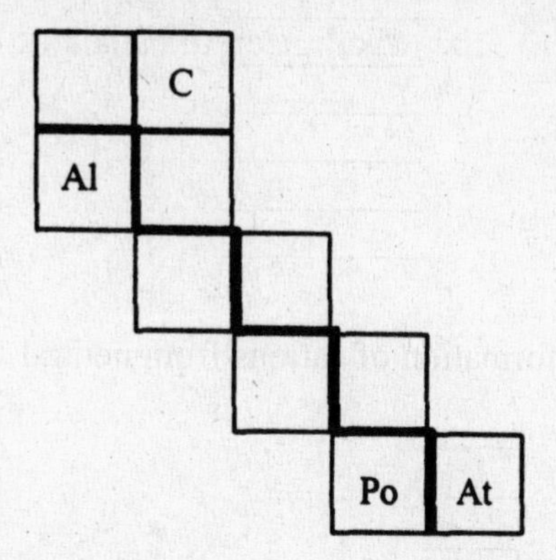

28. Some elements exist in nature as diatomic molecules. Which of the elements below will be found as diatomic molecules?

a. Ar

b. O

c. K

d. F

e. S

f. N

g. H

h. Cl

29. At room temperature, what is the physical state (solid, liquid, or gas) of each of the elements that naturally form diatomic molecules?

30. Which elements are always found in nature as individual atoms?

a. carbon

b. krypton

c. magnesium

d. chlorine

e. helium

f. neon

g. aluminum

h. sulfur

31. Fill in the name of the correct element next to its description at 25°C.

a. Liquid metal ________
b. Yellow green gas ________
c. Colorless gas ________
d. A 2-carat diamond ________
e. A reddish-brown liquid ________
f. Dark purple solid ________

32. Balance the equations for the formation of cations from neutral atoms.

a. Ca → ______ + ______
b. K → ______ + ______
c. Sr → ______ + ______
d. Rb → ______ + ______

33. Balance the equations for the reactions of cations with electrons.

a. Mg^{2+} + ______ → ______
b. Li^{+} + ______ → ______
c. $2H^{+}$ + ______ → ______
d. Na^{+} + ______ → ______

34. Fill in the correct number of protons for either the element or the ion in the table below.

Element	Protons	Electrons	Ion	Protons	Electrons
potassium	19	19	______	______	______
oxygen	______	______	______	8	10
bromine	35				36
strontium	______	38	______	38	
aluminum	13	______	______	______	10

35. You wish to find out whether the compound MgF_2 is composed of ions. What test could you perform to help you make a decision?

36.

a. Which diagram represents a solid NaCl crystal?

b. Which diagram represents NaCl dissolved in water?

c. Which form of NaCl, solid or aqueous solution, allows free movement of ions?

i. 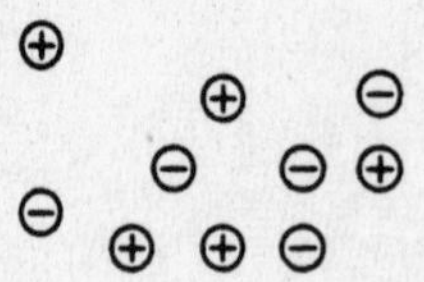 ii.

37. How many of each ion are needed to form a neutral compound?

 a. Ca^{2+} and F^-
 b. Mg^{2+} and O^{2-}
 c. Na^+ and S^{2-}
 d. Li^+ and I^-
 e. Sr^{2+} and Cl^-
 f. K^+ and P^{3-}
 g. Na^+ and N^{3-}
 h. Na^+ and N^{3-}

38. What is wrong with the formulas below? Write the correct formula for each.

 a. $AlCl_2$
 b. NaO
 c. Mg_2P
 d. CaI_3
 e. LiN_3
 f. KS

ANSWERS TO LEARNING REVIEW

1.

 a. bromine
 b. Group 7
 c. halogens
 d. Br_2
 e. liquid
 f. bromide ion, Br^-
 g. 45

2. Silicon, aluminum and iron are found in large amounts on earth, but in small amounts in the human body.

3. Note that some of the elements are found in more than one category.

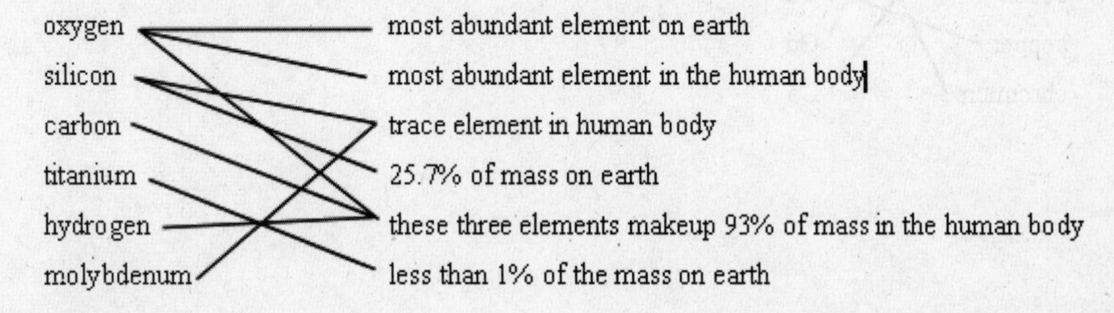

4.

a. As

b. F

c. Mg

d. Fe

e. Ne

f. Pb

g. K

h. Cr

i. N

j. Ca

5. Boron, carbon, fluorine, iodine, nitrogen, oxygen, phosphorous, potassium, sulfur, tungsten and uranium all have one-letter symbols.

6.

a. tungsten

b. mercury

c. copper

d. potassium

e. iron

f. lead

g. antimony

h. sodium

7. There are quite a few elements whose symbols begin with the letter "c." The symbols for these elements are therefore similar to each other.

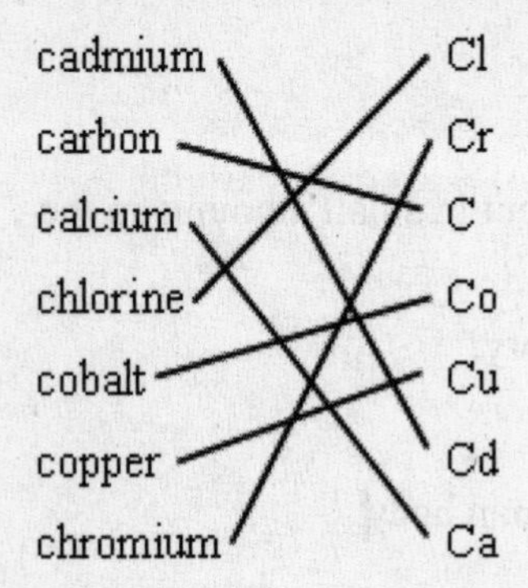

8.

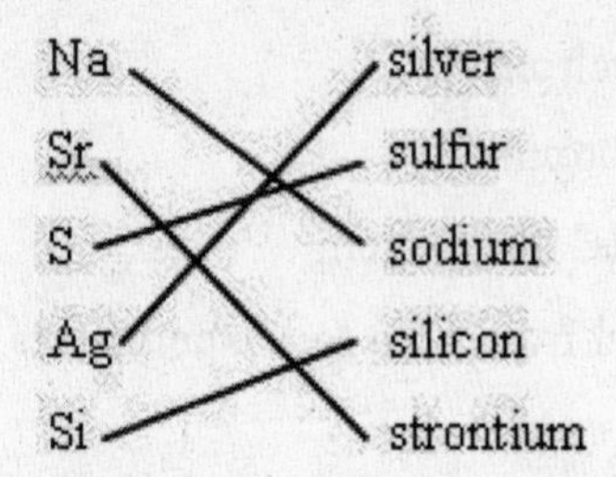

9. Dalton's atomic theory states that all elements are made of atoms. For any one element, all the atoms are the same (Dalton didn't know about isotopes). Different elements are made from different kinds of atoms. Atoms from different elements can combine to make compounds. Each compound always has the same relative numbers and kinds of atoms. Chemical reactions do not cause new elements to form.

10. The law of constant composition states that compounds always have the same proportions of each element by mass. Dalton's model states that compounds always have the same relative numbers and kinds of atoms. If compounds always have the same relative number of atoms, they will also have a constant proportion by mass. For example, the compound carbon dioxide always has one carbon atom for two oxygen atoms. The ratio of the mass of a carbon atom to the mass of two oxygen atoms also stays constant for molecules of carbon dioxide. This relationship was predicted by Dalton's model.

11. Dalton's model states that atoms from different elements can combine to produce compounds, and each compound always has the same relative numbers and kinds of atoms. Dalton predicted that different compounds would be found that were made of the same kinds of atoms, but combined in different numbers. The discovery of NO, NO_2, and N_2O confirmed Dalton's prediction and supported his model.

12.

a. C_2H_6O

b. $MgBr_2$

c. P_4O_{10}

d. AsH_3

13. Remember that when an element symbol has no subscript, only one atom of that element is present. Subscript numbers always refer to the element to the *left* of the subscript number.

a. There are three atoms total and three different elements.

b. There are five atoms total and two different elements.

c. There are five atoms total and two different elements.

d. There are four atoms total and two different elements.

e. There are eight atoms total and three different elements.

14. Because the particles were repelled by the negative pole, it was believed that the particles were negatively charged because like charges repel each other. The electron is a subatomic particle with a negative charge, so the correct answer is c.

15.

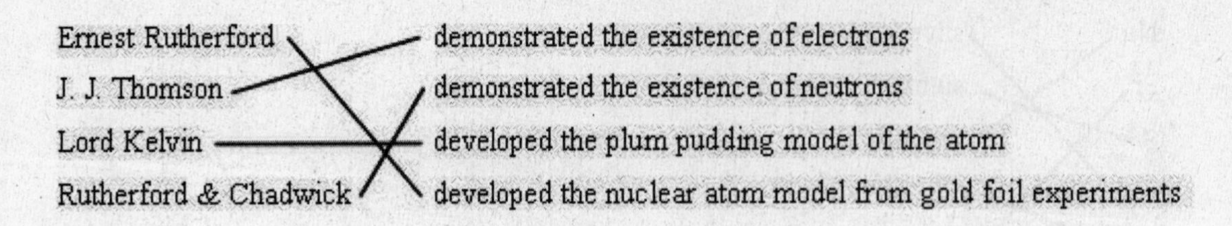

16.

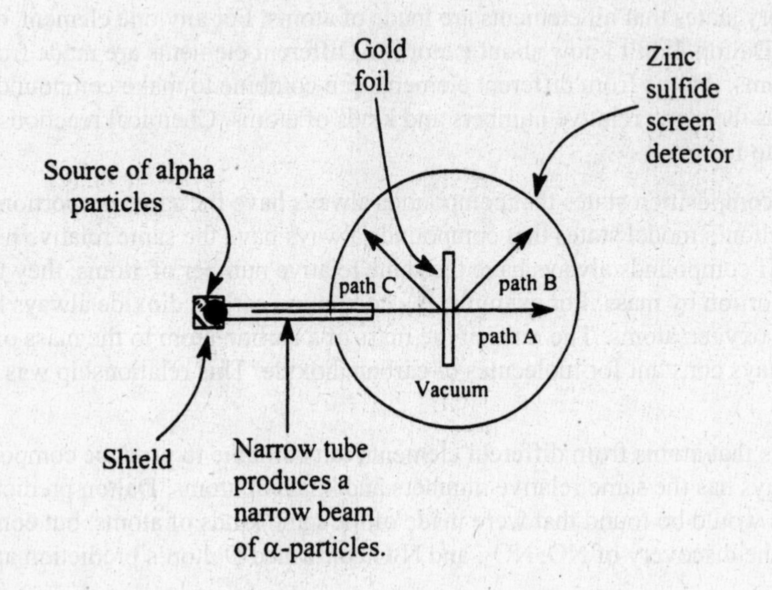

17.

a. An atom consists mostly of empty space.

b. α-particles have a positive charge because they contain two protons. When a moving α-particle travels close to the nucleus of an atom that itself contains protons, the α-particle is deflected from its path because two areas of positive charge repel each other.

c. Some of the α-particles scored a direct hit and bounced straight back. The particle that the α-particle hit must be an area within the atom that is very massive for the heavy α-particle to bounce straight back.

18.

	Relative Mass	Relative Charge
a. Electron	1	1 −
b. Proton	1836	1 +
c. Neutron	1839	0

19. Isotopes of all atoms have the same number of protons as they do electrons. All sodium isotopes have 11 protons and 11 electrons. So the answer is false.

20.

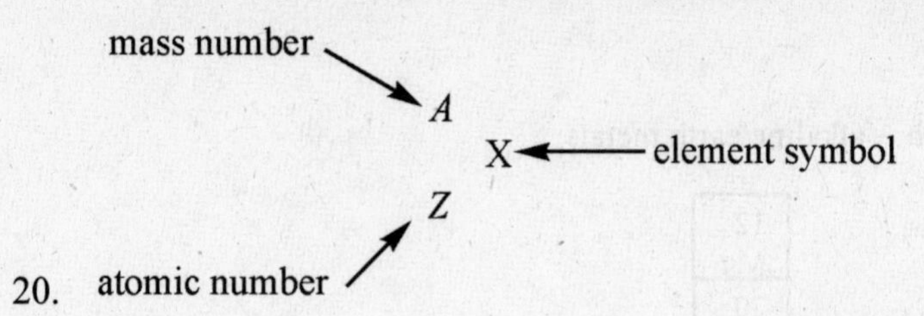

21.

a. $^{3}_{1}\text{H}$

b. $^{37}_{17}\text{Cl}$

c. $^{18}_{8}\text{O}$

d. $^{235}_{92}\text{U}$

e. $^{32}_{16}\text{S}$

22.

a. The mass number provides the number of protons plus the number of neutrons. If the number of neutrons is 24, then the number of protons is 46 minus 24, or 22 protons.

b. The number of protons always equals the number of electrons, so there are 22 electrons in this isotope.

23.

a. When isotopes are designated with the element name followed by a number, as in aluminum-29, the number is e

b. Aluminum-29 has 16 neutrons.

24.

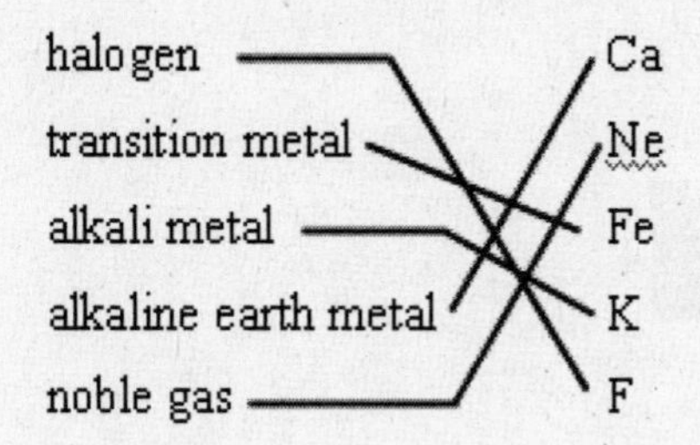

25.

a. halogens

9 F
17 Cl
35 Br
53 I

b. alkaline earth metals

12 Mg
20 Ca
38 Sr
56 Ba

c. noble gases

2 He
10 Ne
18 Ar
36 Kr
54 Xe

d. alkali metals

3 Li
11 Na
19 K
37 Rb

26. C, P, Br, and I are nonmetals. The nonmetals are found to the right of the jagged line on the right side of the periodic table.

27.

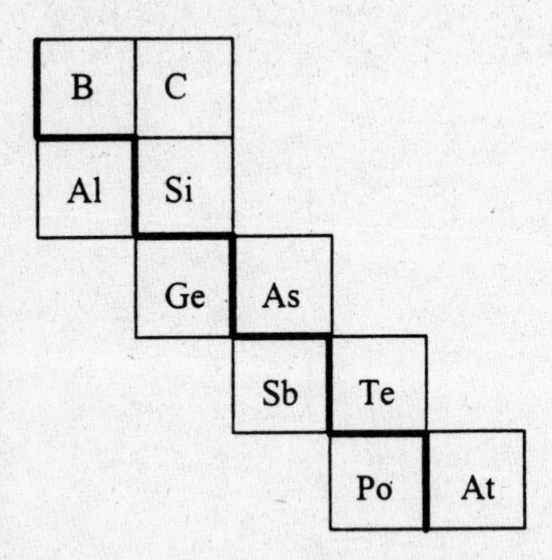

28.

a. Oxygen is found in nature as O_2 molecules.

d. Fluorine is found in nature as F_2 molecules.

f. Nitrogen is found in nature as N_2 molecules.

g. Hydrogen is found in nature as H_2 molecules.

h. Chlorine is found in nature as Cl_2 molecules.

29. Most of the diatomic molecules—H_2, N_2, O_2, F_2, Cl_2—are gases at room temperature. Br_2 is a reddish-brown liquid; I_2 is a dark purple solid.

30.

a. Carbon is usually found combined with other elements such as hydrogen.

b. Krypton is always found as individual krypton atoms.

c. Magnesium is usually found combined with other elements.

d. Chlorine is found as either Cl_2, or combined with other elements.

e. Helium is always found as individual helium atoms.

f. Neon is always found as individual neon atoms.

g. Aluminum is usually found combined with other elements.

h. Sulfur is usually found as S_8 molecules or combined with other elements.

31. Only a few of the elements have unique properties, but there are many elements that could be described as colorless gases or as shiny metals.

a.	Liquid metal	mercury
b.	Yellow green gas	chlorine
c.	Colorless gas	Many elements fit this description, such as oxygen, hydrogen, etc.
d.	A 2 carat diamond	carbon
e.	A reddish-brown liquid	bromine
f.	Dark purple solid	iodine

32. Cations have lost one or more electrons, and the number of electrons lost always equals the charge on the cation.

a. $Ca \rightarrow Ca^{2+} + 2e^-$

b. $K \rightarrow K^+ + e^-$

c. $Sr \rightarrow Sr^{2+} + 2e^-$

d. $Rb \rightarrow Rb^+ + e^-$

33. Cations will react with electrons to form neutral atoms.

a. $Mg^{2+} + 2e^- \rightarrow Mg$

b. $Li^+ + e^- \rightarrow Li$

c. $2H^+ + 2e^- \rightarrow H_2$

d. $Na^+ + e^- \rightarrow Na$

34. The number of protons does not change when neutral atoms form ions, but the number of electrons either increases or decreases.

Element	Protons	Electrons	Ion	Protons	Electrons
potassium	19	19	potassium	19	18
oxygen	8	8	oxide	8	10
bromine	35	35	bromide	35	36
strontium	38	38	strontium	38	36
aluminum	13	13	aluminum	13	10

35. You can place some solid MgF_2 in water. If the solid MgF_2 contains Mg^{2+} ions and F^- ions, when the compound dissolves in water an aqueous solution of Mg^{2+} and F^- ions will form. Test whether or not a solution of MgF_2 will conduct an electrical current by immersing electrodes in the solution. If the solution allows current to flow and a bulb to shine, there is evidence that ions are in solution, free to move around, and able to conduct an electrical current.

36.

a. A solid NaCl crystal is an ordered rigid structure, structure i.

b. Ions are pulled away from the orderly crystal by water molecules, as in structure ii.

c. Ions in water are free to move around and are not packed close together. Those in a solid are ordered and packed close together so that each anion is surrounded by cations.

37.

a.	one calcium and two fluoride ions	CaF_2
b.	one magnesium and one oxide ion	MgO
c.	two sodium and one sulfide ion	Na_2S
d.	one lithium and one iodide ion	LiI
e.	two strontium and two chloride ions	$SrCl_2$
f.	three potassium and one phosphide ion	K_3P
g.	three sodium and one nitride ion	Na_3N

38.

a. Aluminum forms ions with a 3+ charge, and chlorine forms ions with a 1– charge. Three chlorine ions will combine with one aluminum ion to form $AlCl_3$.

b. Sodium forms ions with a 1+ charge, and oxygen forms ions with a 2– charge. Two sodium ions will combine with one oxide to form Na_2O.

c. Magnesium forms ions with a 2+ charge, and phosphorus forms ions with a 3– charge. Three magnesium ions will combine with two phosphide ions to form Mg_3P_2.

d. Calcium forms ions with a 2+ charge, and iodine forms ions with a 1– charge. One calcium ion will combine with two iodide ions to form CaI_2.

e. Lithium forms ions with a 1+ charge, and nitrogen forms ions with a 3– charge. Three lithium ions will combine with one nitride ion to form Li_3N.

f. Potassium forms ions with a 1+ charge, and sulfur forms ions with a 2– charge. Two potassium ions will combine with one sulfide ion to form K_2S.

CHAPTER 5

Nomenclature

INTRODUCTION

In this chapter you will learn how to name ions and compounds. You will be asked to learn rules for naming ions and compounds, and you will need to memorize some names. Many of the problems at the end of this chapter have as their goal the application of the rules you will learn. Work on the naming problems until you have developed skill at naming chemical compounds.

CHAPTER DISCUSSION

Many students look at nomenclature as memorizing the names of a seemingly endless list of chemicals, but this is simply not true. There are systematic rules for naming compounds. By knowing only a few rules you can name most any compound you will encounter in introductory chemistry.

For example, all the charges for the common, simple cations and anions in Table 5.1 of your text come from their placement on the periodic table (see Chapter 4 if this is unfamiliar to you).

Also, look at Table 5.2, the "common type II cations." What is the difference here? The difference is that these ions come from transition metals (not from the alkali or alkaline earth metals in the first two columns of the periodic table). These ions can have more than one stable charge, and we therefore have to specify which ion we are talking about. The Roman numeral is the charge, not the subscript. Therefore, the formula Fe_2O_3 should be read as "iron (III) oxide." Make sure to prove to yourself that it is indeed iron (III); that is, Fe^{3+}.

Notice this is different from NaCl. The name "sodium (I) chloride" is not so much wrong as it is redundant. The most stable charge for the sodium ion in a compound is 1+, and the most stable charge for the chloride ion in a compound is 1−. Therefore the only possible formula for "sodium chloride" is NaCl. However, the name "iron oxide" is incomplete because it could be referring to Fe_2O_3 [iron (III) oxide] or FeO [iron (II) oxide]. We use Roman numerals only when we have to for clarity.

The same goes for prefixes. For example, there is no reason to call $CaCl_2$ "calcium dichloride" because $CaCl_2$ is the only stable formula for an ionic compound made of calcium and chloride ions. Therefore, we can simply say "calcium chloride," and everyone knows that this means "$CaCl_2$." However, we cannot use the name "carbon oxide" because these are both nonmetals, meaning the compound is not made up of ions. There is no way from the name "carbon oxide" to know the formula (although we do know the compound is made from carbon and oxygen). Therefore we have to specify if it is carbon monoxide (CO) or carbon dioxide (CO_2), for example. The only way to make sure you can name compounds is to practice. Being able to name compounds requires you to have some rules memorized, but it also requires you to understand these rules.

LEARNING REVIEW

1. Name the cations and anions below.
   a. Cl^-
   b. Mg^{2+}
   c. Li^+
   d. Ba^{2+}
   e. N^{3-}
   f. O^{2-}
   g. F^-
2. Name the following Type I binary compounds.
   a. KF
   b. CaS
   c. NaI
   d. Li_3N
   e. HCl
   f. Al_2O_3
   g. AgCl
   h. MgF_2
3. The following elements can all form more than one cation. How many cations form, and what is the charge on each of them?
   a. Cu
   b. Fe
   c. Sn
   d. Hg
   e. Pb
4. Name the following Type II binary compounds.
   a. $FeCl_3$
   b. PbO_2
   c. CoI_2
   d. SnF_2
   e. Fe_2S_3
   f. Hg_2Br_2

5. Name each of the compounds below.

   a. $CaBr_2$

   b. PbS

   c. AlP

   d. FeS

   e. CoO

   f. $MgCl_2$

6. Each of the compounds below has an *incorrect* name. Name each one correctly.

   a. KBr potassium(I) bromide
   b. Cu_2O cupric oxide
   c. PbS_2 lead(IV) sulfide(II)
   d. Na_3P sodium(III) phosphide
   e. $FeCl_3$ iron chloride

7. Name the following Type III binary compounds.

   a. PCl_5

   b. CCl_4

   c. N_2O_3

   d. S_2F_{10}

   e. SO_2

   f. CO

8. Name the following Type I, Type II, or Type III compounds.

   a. KI

   b. NO_2

   c. $FeCl_2$

   d. Al_2O_3

   e. Cl_2O_7

   f. CaS

9. What are the names of the polyatomic ions below?

   a. HCO_3^-

   b. OH^-

   c. NH_4^+

   d. NO_2^-

   e. SO_4^{2-}

   f. CrO_4^{2-}

10. The compounds below all contain polyatomic ions. Name each one.

    a. K_2SO_4

    b. $Fe(OH)_3$

    c. NH_4NO_3

    d. $Al_2(Cr_2O_7)_3$

    e. $Ca(CN)_2$

    f. $Mg_2(PO_4)_2$

    g. $NaMnO_4$

    h. $Cu(ClO_3)_2$

    i. $PbCO_3$

11. Check your knowledge of the common acids by naming the acids below.

    a. H_2SO_4

    b. HCN

    c. HBr

    d. HNO_3

    e. H_2S

    f. $HC_2H_3O_2$

12. From their names, write formulas for the compounds below.

    a. aluminum chloride

    b. cobalt(III) permanganate

    c. dinitrogen trioxide

    d. sulfur dioxide

    e. calcium nitrate

    f. silver chloride

    g. iron(II) acetate

    h. tin(IV) chlorite

    i. sodium sulfate

    j. lithium hydrogen carbonate

    k. mercury(II) dichromate

ANSWERS TO LEARNING REVIEW

1. The ions in this problem are all monatomic ions. The cations all have the same name as the element while the anions all end in –ide.

    a. chloride

    b. magnesium

    c. lithium

d. barium

e. nitride

f. oxide

g. fluoride

2. Type I binary compounds form between a metal and a nonmetal.

a. potassium fluoride

b. calcium sulfide

c. sodium iodide

d. lithium nitride

e. hydrogen chloride

f. aluminum oxide

g. silver chloride

h. magnesium fluoride

3. Notice that these are all transition metal ions.

a. Cu^{+} Cu^{2+}
b. Fe^{2+} Fe^{3+}
c. Sn^{2+} Sn^{4+}
d. Hg_2^{2+} Hg^{2+}
e. Pb^{2+} Pb^{4+}

4. Type II binary compounds form between a metal that forms more than one cation and a nonmetal.

a. iron(III) chloride

b. lead(IV) oxide

c. cobalt(II) iodide

d. tin(II) iodide

e. iron(III) sulfide

f. mercury(I) bromide

5. The compounds are mixed Type I and Type II binary compounds.

a. calcium bromide

b. lead(II) sulfide

c. aluminum phosphide

d. iron(II) sulfide

e. cobalt(II) oxide

f. magnesium chloride

6.

a. Potassium forms only cations with 1+ charge, so potassium(I) bromide should be potassium bromide.

b. The formula Cu_2O shows copper with a 1+ charge, which is named the copper(I) or cuprous ion. The correct name for this formula is copper(I) oxide or cuprous oxide.

c. The formula PbS_2 tells us that the charge on the lead ion is 4+, so the first part of the name, lead(IV), is correct. The sulfide ion has a 2– charge, but we do not use Roman numerals after the anion name. So the correct name for this compound is lead(IV) sulfide.

d. Sodium forms only cations with 1+ charge, so sodium(III) phosphide should be sodium phosphide.

e. The formula $FeCl_3$ tells us that iron has a 3+ charge. Because iron forms cations with more than one charge, the correct name would be iron(III) chloride.

7. Type III binary compounds form between nonmetals. The prefix that indicates ten atoms is *deca-*. This prefix is used in problem 7d.

a. phosphorus pentachloride

b. carbon tetrachloride

c. dinitrogen trioxide

d. disulfur decafluoride

e. sulfur dioxide

f. carbon monoxide

8. The compounds are a mixture of Type I, Type II and Type III compounds.

a. potassium iodide

b. nitrogen dioxide

c. iron(II) chloride

d. aluminum oxide

e. dichlorine heptoxide

f. calcium sulfide

9. If you have trouble naming these ions, go back and review the names again. You will need these names throughout your chemistry career.

a. bicarbonate

b. hydroxide

c. ammonium

d. nitrite

e. sulfate

f. chromate

10.

a. potassium sulfate

b. iron(III) hydroxide

c. ammonium nitrate

d. aluminum dichromate

e. calcium cyanide

f. magnesium phosphate

g. sodium permanganate

h. copper(II) chlorate

i. lead(II) carbonate

11.

a. sulfuric acid

b. hydrocyanic acid

c. hydrobromic acid

d. nitric acid

e. hydrosulfuric acid

f. acetic acid

12.

a. $AlCl_3$

b. $Co(MnO_4)_3$

c. N_2O_3

d. SO_2

e. $Ca(NO_3)_2$

f. $AgCl$

g. $Fe(C_2H_3O_2)_2$

h. $Sn(ClO_2)_4$

i. Na_2SO_4

j. $LiHCO_3$

k. $HgCr_2O_7$

CHAPTER 6

Chemical Reactions: An Introduction

INTRODUCTION

Knowing how to write chemical equations and how to interpret what they mean is an important part of chemistry. If you have learned the symbols for the elements and can write the formulas of compounds from their names, then learning to write chemical equations will be easier. The Answers to Learning Review (#5) discusses the logic used when balancing a chemical equation by trial and error. If you are having trouble balancing equations, go over the steps used to balance the equations in the Answers to Learning Review.

CHAPTER DISCUSSION

You have already seen a molecular level representation of a balanced chemical equation in Chapter 4 (see Figure 4.15). Another such representation of a different chemical equation can be seen in Figure 6.4.

Remember that the point of a chemical equation is to use symbols to show what happens during the chemical reaction. Therefore, while you are first learning to balance chemical equations it is a good idea to think at a molecular level to make sure you are balancing the equation and not changing the composition of a reactant or a product. Learn to relate the words to the representation (molecular level perspective) to the symbols (the coefficients, subscripts, and atomic symbols).

For example, consider the statement,

"Hydrogen gas reacts with oxygen gas to produce water vapor."

We would like to write this reaction as an equation in terms of the symbols for the elements. What is the advantage to this? First, it is generally easier to write once you understand the language. But more importantly, using the symbols allows us to balance the equation, which means we are able to determine the relative amounts of the reactants we need (in this case hydrogen gas and oxygen gas) along with the relative amount of product produced (in this case, the water).

Let's look at the reaction statement in terms of a molecular-level sketch. To do this, we have to know the formulas for hydrogen gas, oxygen gas, and water and how to sketch them. Recall that hydrogen and oxygen gases are diatomic (two atoms per molecule) and that the water molecule consists of two hydrogen atoms and one oxygen atom. Thus, we can sketch a representation of the reaction as follows:

While this conveys what occurs, it is not balanced. That is, it does not give us information about the relative amounts of reactants and products. Notice, for example, that there are two oxygen atoms on the left side of the equation and only one oxygen atom on the right side. This simply cannot occur during a chemical reaction.

Because oxygen gas is diatomic, and water consists of only one oxygen atom, a diatomic oxygen molecule can produce two molecules of water.

However, one hydrogen molecule can produce only one molecule of water. To produce two molecules of water would require two molecules of hydrogen gas.

This equation is now balanced. All atoms are accounted for on each side of the equation; that is, all atoms are conserved.

Now we are ready to think about this sketch in terms of chemical symbols. We have four atoms of hydrogen and two atoms of oxygen on each side of the equation, but we also want to convey that hydrogen gas is reacting with oxygen gas to form water. In fact, now that we have balanced our equation, we can be more specific about the amounts and state the reaction as,

> "Two molecules of hydrogen gas react with one molecule of oxygen gas to produce two molecules of water."

We can symbolize the diatomic hydrogen gas as H_2, the diatomic oxygen gas as O_2, and water as H_2O. We have balanced the equation with our molecular level sketches, so now we can add these numbers to our equation to get

$$2H_2\,(g) + O_2\,(g) \rightarrow 2H_2O\,(l).$$

Note there is a 2:1:2 ratio of molecules just as we determined with the molecular-level sketches (we generally do not include the "1" in front of a molecule–it is assumed). These numbers are called the coefficients and represent the ratio of molecules (reactants and products) in the equation. Recall from Chapter 4 that the subscripts in the molecules tell us how many atoms of a particular element are in one molecule.

One common mistake made when first balancing equations is to change subscripts. For example, if we look at the unbalanced equation,

$$H_2\,(g) + O_2\,(g) \rightarrow H_2O\,(l),$$

it may seem reasonable to balance this equation by adding another oxygen atom to the water molecule to get

$$2H_2\,(g) + O_2\,(g) \rightarrow 2H_2O_2\,(l).$$

What is wrong with this? Think back to our molecular level drawings and how you would sketch H_2O compared to H_2O_2. They are different molecules with different chemical properties. Changing the subscript changes the identity of the chemical. In this case, for example, H_2O_2 is hydrogen peroxide which is quite different from water. The goal to balancing a chemical equation is not to just make sure there are the same numbers of each type of atom on both sides of the equation, but to balance the equation that is given to you. In this case we wanted to balance the equation that represented the production of water, not hydrogen peroxide.

For another example, answer the following question (think about it before reading on).

The reaction of an element X (Δ) with element Y (O) is represented in the following diagram. Which of the equations best describes this reaction?

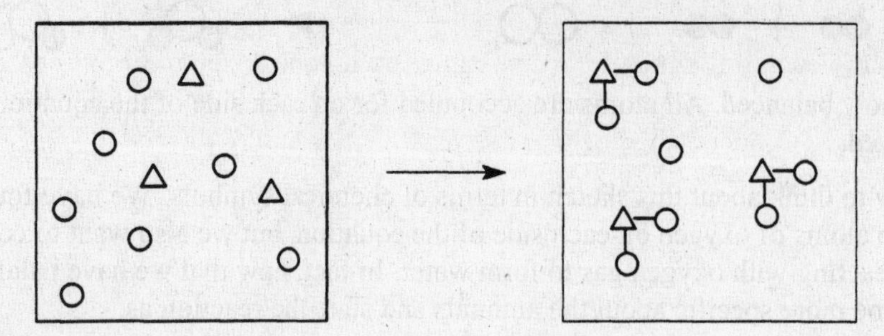

a. $3X + 8Y \rightarrow X_3Y_8$

b. $3X + 6Y \rightarrow X_3Y_6$

c. $X + 2Y \rightarrow XY_2$

d. $3 + 8Y \rightarrow 3XY_2 + 2Y$

e. $X + 4Y \rightarrow XY_2$

The correct answer is "c". Choices "a" and "b" give the wrong products. The product formed should be symbolized as "XY_2" from the molecular drawing. Choices "c," "d," and "e" have this as a product, but choice "e" is not balanced. So what is wrong with choice "d"? Many students choose this because it correctly gives the number of X's and Y's on the reactant side and shows that there are two Y's left over when the reaction is completed. So why is it incorrect?

First of all, the chemicals on the right side of the equation are the products, that is, they are produced in the reaction. If choice "d" is correct, this implies that Y's were produced in this reaction, and this is not true; that is, Y's may be left over, but they were not made. Also, the fact that we started with three X's and eight Y's is not relevant in this case. The balanced equation does not (repeat, does **not**) tell us how much of each chemical we have, but it gives us a ratio of the amounts that react or are produced. Think of it in terms of a recipe. A recipe is written to tell you how much of each ingredient is needed to react with the others in order to make a certain amount of product. It does not tell you how much of each ingredient you happen to have in your kitchen. Thus, choice "c" tells us what is actually reacting; that is, for every one X, two Y's react to form one XY_2.

Could we symbolize the reaction the following way?

$$2X + 4Y \rightarrow 2XY_2$$

Yes we could. However, we generally reduce all of the coefficients to the least common whole numbers for the sake of simplicity. This brings up an important point; the value of an individual coefficient is not important. What is important is the ratio of the coefficients.

Suppose, for example, you want a recipe for chocolate chip cookies, and you are told that you will need flour, sugar, eggs, baking soda, salt, and one egg. Telling you that you need one egg is not very helpful since you don't know the amounts of the other ingredients nor do you know how many cookies it will make. And realize that even if you are told all the amounts, you can change them to make the number of cookies you want. Thus the actual numbers are not as important as the ratio.

For example, recall the balanced equation

$$2H_2 + O_2 \rightarrow 2H_2O$$

This equation can be read as,

"Two molecules of diatomic hydrogen react with one molecule of diatomic oxygen to produce two molecules of water"

or

"Two dozen molecules of diatomic hydrogen react with one dozen molecules of diatomic oxygen to produce two dozen molecules of water"

or

"Two hundred molecules of diatomic hydrogen react with 100 molecules of diatomic oxygen to produce 200 molecules of water."

In fact, there are an infinite number of possibilities as long as the ratio of the number of molecules of H_2, O_2 and H_2O is 2:1:2, respectively.

Again, the individual coefficient is not important. It is the ratio between reactants and products that is important.

Finally, while you are first balancing equations you may wish to use molecular-level drawings, but you should eventually work toward writing an equation (and then balancing it) using symbols directly from the words. There are many examples of these types of questions at the end of Chapter 6 in your text.

LEARNING REVIEW

1. Which of the following indicates that a chemical reaction has occurred?
   a. Liquid water boils to produce steam.
   b. Burning firewood gives off heat.
   c. Mixing two colorless liquids produces a bright yellow solid.
   d. Solid $NaHCO_3$ dissolves in water.
2. Why is it important that chemical equations be balanced?
3. Count the number of each kind of atom on both sides of the equation, and decide which reactions are balanced and which are not.
   a. $H_2 + Br_2 \rightarrow HBr$
   b. $KClO_3 \rightarrow KCl + O_2$
   c. $2NaOH + CO_2 \rightarrow Na_2CO_3 + H_2O$
   d. $C_2H_5OH + 3O_2 \rightarrow 2CO_2 + 3H_2O$
   e. $3Cu + HNO_3 \rightarrow 3Cu(NO_3)_2 + NO + H_2O$
4. Use the following word descriptions to write unbalanced chemical equations showing the formulas of reactants and products. Make sure you include the physical states of reactants and products.
   a. Solid iron metal reacts with oxygen in the atmosphere to form rust, iron(III) oxide.
   b. Solid magnesium metal reacts with aqueous hydrochloric acid to produce hydrogen gas and an aqueous solution of magnesium chloride.
   c. Solid silver oxide decomposes upon heating to produce solid silver metal and oxygen gas.
   d. Aqueous sodium hydroxide reacts with aqueous nitric acid to produce aqueous sodium nitrate and liquid water.

5. Balance these chemical equations. Check your work by counting the number of each kind of atom on both sides of the equation.

   a. $KOH(aq) + H_2S(aq) \rightarrow K_2S(aq) + H_2O(l)$

   b. $HNO_2(aq) \rightarrow N_2O_3(g) + H_2O(aq)$

   c. $NaOH(aq) + H_2SO_4(aq) \rightarrow Na_2SO_4(aq) + H_2O(l)$

   d. $(NH_4)_2S(aq) + Pb(NO_3)_2(aq) \rightarrow PbS(s) + NH_4NO_3(aq)$

   e. $Al(s) + O_2(g) \rightarrow Al_2O_3(s)$

6. Balance these chemical equations.

   a. $SO_2(g) + O_2(g) \rightarrow SO_3(g)$

   b. $C_4H_{10}(g) + O_2(g) \rightarrow CO_2(g) + H_2O(g)$

   c. $Fe_2O_3(s) + C(s) \rightarrow Fe(s) + CO_2(g)$

   d. $TiCl_4(l) + H_2O(l) \rightarrow TiO_2(s) + HCl(aq)$

7. Balance these chemical equations.

   a. $KI(aq) + Br_2(l) \rightarrow KBr(aq) + I_2(s)$

   b. $PbO_2(s) \rightarrow PbO(s) + O_2(g)$

   c. $Fe(OH)_3(s) + H_2SO_4(aq) \rightarrow Fe_2(SO_4)_3(s) + H_2O(l)$

   d. $K_3PO_4(aq) + BaCl_2(aq) \rightarrow KCl(aq) + Ba_3(PO_4)_2(s)$

ANSWERS TO LEARNING REVIEW

1.

   a. Boiling represents a physical change; not a chemical reaction.

   b. Heat production is an indication of a chemical reaction.

   c. A color change and the production of a solid (a new substance) are both indications of a chemical reaction.

   d. No chemical reaction has occurred. Solid $NaHCO_3$ dissolves into ions in water. The ions are so small they cannot be seen, but the identity of the compound does not change.

2. When a chemical reaction occurs, atoms are neither created nor destroyed. The atoms are only rearranged to produce new molecules. Therefore it is important that the same kinds and numbers of atoms be present on both sides of a chemical equation.

$H_2 + Br_2$	$\rightarrow$	HBr	
2 H		1 H	
2 Br		1 Br	unbalanced

3.

   a.

$KClO_3$	$\rightarrow$	$KCl + O_2$	
1 K		1 K	
1 Cl		1 Cl	
3 O		2 O	unbalanced

b.

$2NaOH + CO_2$	$\rightarrow$	$Na_2CO_3 + H_2O$	
2 Na		2 Na	
2 O		1 C	
2 H		3 O	
1 C		2 H	
2 O		1 O	
The **total** number of atoms of each kind is			
2 Na		2 Na	
4 O		4 O	
2 H		2 H	
1 C		1 C	balanced

c.

$C_2H_5OH + 3O_2$	$\rightarrow$	$2CO_2 + 3H_2O$	
2 C		2 C	
(5 + 1) H		(2 × 2) O	
1 O		(3 × 2) H	
(3 × 2) O		3 O	
The **total** number of atoms of each kind is			
2 C		2 C	
6 H		6 H	
7 O		7 O	balanced

d.

$3Cu + HNO_3$	$\rightarrow$	$3Cu(NO_3)_2 +$ $NO + H_2O$	
3 Cu		3 Cu	
1 H		(3 × 2) O	
1 N		2 N	
3 O		1 N	
		1 O	
		2 H	
		1 O	
The **total** number of atoms of each kind is			
3 Cu		3 Cu	
1 H		2 H	
1 N		3 N	
3 O		8 O	unbalanced

4. This problem requires that you be able to write formulas from word descriptions. Do not forget to include the physical state of both products and reactants.

   a. $Fe(s) + O_2(g) \rightarrow Fe_2O_3(s)$

   b. $Mg(s) + HCl(aq) \rightarrow H_2(g) + MgCl_2(aq)$

   c. $Ag_2O(s) \rightarrow Ag(s) + O_2(g)$

   d. $NaOH(aq) + HNO_3(aq) \rightarrow NaNO_3(aq) + H_2O(l)$

5.

a. First find the most complex formula. KOH contains three different kinds of atoms, so begin by adjusting the coefficients of the atoms in KOH. There are two potassium atoms on the right and only one on the left. Adjust potassium by increasing the coefficient of KOH from 1 to 2.

$$2KOH(aq) + H_2S(aq) \rightarrow K_2S(aq) + H_2O(l)$$

Left	Right
2 K	2 K
2 O	1 O
4 H	2 H
1 S	1 S

Now, K and S are balanced, but oxygen and hydrogen are not. There are four hydrogen and two oxygen atoms on the left, but only two hydrogen and two oxygen atoms on the right. If we adjust the coefficient of water to 2, oxygen and hydrogen are the same on each side, and the equation is balanced.

$$2KOH(aq) + H_2S(aq) \rightarrow K_2S(aq) + 2H_2O(l)$$

Left	Right
2 K	2 K
2 O	2 O
4 H	4 H
1 S	1 S

b. Because HNO_2 is the most complex molecule, begin by adjusting the number of nitrogen atoms on both sides of the equation. There are two on the right, but only one on the left. Increase the coefficient of HNO_2 to 2. Hydrogen and oxygen are the same on each side, and the equation is balanced.

$$2HNO_2(aq) \rightarrow N_2O_3(g) + H_2O(aq)$$

Left	Right
2 H	2 H
2 N	2 N
4 O	4 O

c. Either NaOH or H_2SO_4 is a good place to begin balancing. Let's start by adjusting the number of sodium ions. Put a coefficient of 2 in front of NaOH.

$$2NaOH(aq) + H_2SO_4(aq) \rightarrow Na_2SO_4(aq) + H_2O(l)$$

Left	Right
2 Na	2 Na
6 O	5 O
4 H	2 H
1 S	1 S

There is one sulfur atom on each side, but the left side now has four hydrogen and six oxygen atoms while the right side has only five oxygen atoms and two hydrogen atoms. The right side can be adjusted by placing a coefficient of 2 in front of H_2O. The equation is now balanced.

$$2NaOH(aq) + H_2SO_4(aq) \rightarrow Na_2SO_4(aq) + 2H_2O(l)$$

Left	Right
2 Na	2 Na
6 O	6 O
4 H	4 H
1 S	1 S

d. Begin by adjusting the ammonium ion and the nitrate ion by putting a coefficient of 2 in front of $NH_4NO_3(aq)$. The equation is now balanced.

$$(NH_4)_2S(aq) + Pb(NO_3)_2(aq) \rightarrow PbS(s) + 2\,NH_4NO_3(aq)$$

Left	Right
4 N	4 N
8 H	8 H
1 S	1 S
1 Pb	1 Pb
6 O	6 O

e. The numbers of aluminum atoms and oxygen atoms on the left are less than on the right. Begin by increasing the coefficient of aluminum to 2. Aluminum is now balanced on both sides.

$$2Al(s) + O_2(g) \rightarrow Al_2O_3(s)$$

Left	Right
2 Al	2 Al
2 O	3 O

Aluminum is balanced, but oxygen is not. There are three oxygen atoms on the right and two on the left. We cannot use a coefficient on the left to produce three oxygen atoms so that the left and right balance. But we can put a coefficient of 2 in front of Al_2O_3 to make six oxygen atoms. A coefficient of 3 on the left adjusts the oxygen atoms on the left to six.

$$2Al(s) + 3O_2(g) \rightarrow 2Al_2O_3(s)$$

Left	Right
2 Al	4 Al
6 O	6 O

There are now four aluminum atoms on the right and two on the left, so increase the coefficient of aluminum on the left to 4. The equation is now balanced.

$$4Al(s) + 3O_2(g) \rightarrow 2Al_2O_3(s)$$

Left	Right
4 Al	4 Al
6 O	6 O

6.

a. $2SO_2(g) + O_2(g) \rightarrow 2SO_3(g)$

b. $2C_4H_{10}(g) + 13O_2(g) \rightarrow 8CO_2(g) + 10H_2O(g)$

c. $2Fe_2O_3(s) + 3C(s) \rightarrow 4Fe(s) + 3CO_2(g)$

d. $TiCl_4(l) + 2H_2O(l) \rightarrow TiO_2(s) + 4HCl(aq)$

7.

a. $2KI(aq) + Br_2(l) \rightarrow 2KBr(aq) + I_2(s)$

b. $2PbO_2(s) \rightarrow 2PbO(s) + O_2(g)$

c. $2Fe(OH)_3(s) + 3H_2SO_4(aq) \rightarrow Fe_2(SO_4)_3(s) + 6H_2O(l)$

d. $2K_3PO_4(aq) + 3BaCl_2(aq) \rightarrow 6KCl(aq) + Ba_3(PO_4)_2(s)$

CHAPTER 7

Reactions in Aqueous Solutions

INTRODUCTION

Water is a good solvent in the chemical laboratory because many compounds are soluble in water. Many chemical reactions occur in water, and you will study some of them in this chapter. To predict what will happen when two compounds that are dissolved in water are mixed, you will need to learn some rules. These rules will help you predict what kinds of compounds are soluble in water and what kinds of products you can make.

Water is also an important part of our environment, and many common reactions take place in water. For example, the rusting of iron takes place in water. For these reasons, a whole chapter is devoted to what happens when substances are added to water.

There are many different chemical reactions. When you first look at the equation for a reaction, it often looks completely new and unfamiliar. After you learn the material in this chapter, you will be able to classify many of the new reactions you come across into one or more basic categories. Categorizing a reaction is important. By fitting a reaction into a specific familiar category, you automatically know some things about that reaction even if you have never seen that specific reaction before.

CHAPTER DISCUSSION

Precipitation Reactions

When you mix an aqueous solution of lead(II) nitrate with an aqueous solution of potassium iodide (both solutions are colorless), beautiful yellow crystals are produced in a clear solution. What is the formula for these crystals? What is occurring in this reaction? This type of reaction is called a precipitation reaction, and the solid is termed the precipitate. Let's think about what is going on.

By now, you should be able to write the left side of the chemical equation (the reactant side) from the names given in the above paragraph. Try to do this before reading on.

Since lead(II) has a 2+ charge, and nitrate has a 1− charge, the formula for lead(II) nitrate is $Pb(NO_3)_2$. The potassium ion has a charge of 1+, and the iodide ion has a charge of 1−. So the formula for potassium iodide is KI. Thus the reactant side of the chemical equation is

$$Pb(NO_3)_2(aq) + KI(aq) \rightarrow$$

But what are the products? To understand this, let's think about what the reactants "look like." The reactants are both aqueous solutions of ionic compounds (also known as salts). An aqueous solution of an ionic compound will consist of the ions floating separately in solution. Thus we can visualize the reactants as

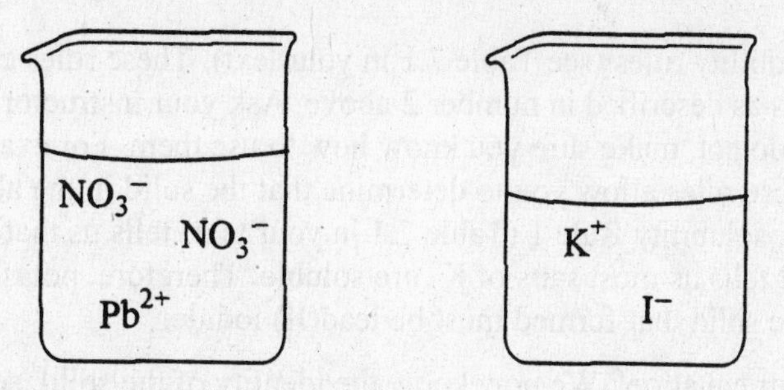

When these solutions are mixed together, the four ions

$$Pb^{2+}, NO_3^-, K^+, I^-$$

are all in solution. Therefore there are four possibilities for products. Try to figure the formulas for these before reading on.

Recall that ions of opposite charges will have attraction for one another and that a molecule will be neutral overall. Therefore we have the following four possibilities for products:

$$Pb(NO_3)_2, KI, PbI_2, KNO_3.$$

We can eliminate the first two possibilities listed above because they are the original reactants. That is, since we know these exist as ions in solution, neither of these two will "reform" simply by mixing with the other solution. Therefore, we know that the possible products are the last two possibilities, and we can write the equation as

$$Pb(NO_3)_2(aq) + KI\ (aq) \rightarrow PbI_2 + KNO_3.$$

As a side note, notice that the subscripts are not necessarily the same on each side. For example, many students make the mistake of writing this equation as (WARNING–THE FOLLOWING CHEMICAL EQUATION IS INCORRECT!)

$$Pb(NO_3)_2(aq) + KI\ (aq) \rightarrow PbI + K(NO_3)_2.$$

Make sure you understand why this is NOT correct. There is no reason that just because a reactant consists of two nitrates (for example) the product must also. This is one reason why it is a good idea to think about the reactants as separate ions when balancing a precipitation reaction.

The correct balanced equation is

$$Pb(NO_3)_2(aq) + 2KI\ (aq) \rightarrow PbI_2 + 2KNO_3$$

This is known as a molecular equation because it shows the complete formulas for all the reactants and products. Note that in this case we have not included phases for the products because we are still not sure which is the solid.

How can we tell which of the two products is the solid? There are a few ways of doing this.

1. We can know something about the possible products. For example, by mixing the solutions a yellow solid was noted. A chemist would know (or could look up) properties of lead(II) iodide and potassium nitrate to see if either was an insoluble yellow solid. In this case, lead(II) iodide is a yellow insoluble solid.

2. We could experiment some more. For example, suppose we mix an aqueous solution of sodium nitrate with an aqueous solution of potassium chloride. In this case, the possible products would be sodium chloride and potassium nitrate (make sure you understand why). However, upon mixing these there is no reaction. Therefore, we know potassium nitrate is not a yellow insoluble solid.

3. We can use the solubility rules (see Table 7.1 in your text). These rules came about by performing similar experiments as described in number 2 above. Ask your instructor if you need to memorize these. Even if you do not, make sure you know how to use them. For example, make sure you understand how these rules allow you to determine that the solid in the above example is lead(II) iodide. In this case, solubility Rule 1 (Table 7.1 in your text) tells us that most nitrate salts are soluble, and Rule 2 tells us most salts of K^+ are soluble. Therefore, potassium nitrate is expected to be soluble, so the solid that formed must be lead(II) iodide.

So, where are we with our equation? We now know the identity of the solid, so we can write

$$Pb(NO_3)_2(aq) + 2KI(aq) \rightarrow PbI_2(s) + 2KNO_3(aq)$$

Again, this is the molecular equation, but in this case we have noted the phase of each reactant and product. However, there are other ways we can represent this equation.

Recall that by writing KI(*aq*), for example, we are stating that the potassium iodide exists in solution as potassium and iodide ions. Thus, we can write the equation as

$$Pb^{2+}(aq) + 2NO_3^-(aq) + 2K^+(aq) + 2I^-(aq) \rightarrow PbI_2(s) + 2K^+(aq) + 2NO_3^-(aq)$$

This more clearly conveys what is occurring in solution. Note that the subscripts that told us how many of a certain ion was present are now coefficients; that is we write

$$2NO_3^-(aq)$$

not

$$(NO_3)_2^-(aq).$$

Also note that the solid is written as a molecule. This is called the complete ionic equation because it contains all the ions that are in solution. This is a rather long way to represent the equation, however, and note that not all of the ions participate. That is, the potassium and nitrate ions do not "do" anything in the reaction (they are termed spectator ions). To simplify this, we can consider only the species that are actually involved in the reaction, and thus we write

$$Pb^{2+}(aq) + 2I^-(aq) \rightarrow PbI_2(s)$$

This is called the net ionic equation. Your text contains many other examples of this in Sections 7.2 and 7.3.

Here are some problems you should be able to answer:

1. Use molecular-level drawings to show what is meant by the terms "strong electrolyte," "insoluble," and "precipitation reaction."

2. If spectator ions do not participate in the reaction, why are they in solution?

3. Mixing an aqueous solution of sodium chloride with an aqueous solution of potassium nitrate is not a chemical reaction. Why not?

Acid–Base Reactions

The acid–base reactions considered in the text are similar to precipitation reactions in that the products can be determined by switching ions. The acid–base reactions are simpler, however, because in all cases that we will consider, one of the products is water.

For example, if we mix hydrochloric acid with an aqueous solution of sodium hydroxide, we get sodium chloride and water as shown in the equation

$$HCl(aq) + NaOH(aq) \rightarrow NaCl(aq) + H_2O(l)$$

The complete ionic equation is

$$H^+(aq) + Cl^-(aq) + Na^+(aq) + OH^-(aq) \rightarrow Na^+(aq) + Cl^-(aq) + H_2O(l)$$

And the net ionic equation is

$$H^+(aq) + OH^-(aq) \rightarrow H_2O(l)$$

This is the net ionic equation for all acid–base reactions we will consider.

Classifying Chemical Reactions

Sections 7.6-7.7 provide an excellent discussion of classifying chemical reactions. Make sense of Figure 7.12. For example, why are combustion, synthesis, and decomposition reactions all oxidation–reduction reactions? Realize that although there are a seemingly infinite number of actual reactions, there are only a few types of reactions. This is similar to nomenclature; by having a systematic approach to looking at chemical reactions we can know a lot about them without an overabundance of memorization.

LEARNING REVIEW

1. Which one of the following does **not** tend to drive a reaction to produce products?
   a. formation of a gas
   b. transfer of electrons
   c. color change
   d. formation of water
   e. formation of a solid
2. Write the formulas for the ions that are formed when these ionic compounds are dissolved in water. How many of each kind of ion are produced for each molecule dissolved?
   a. $(NH_4)_2SO_4$
   b. KNO_3
   c. $Na_2Cr_2O_7$
   d. $MgCl_2$
   e. Li_3PO_4
   f. $Al(NO_3)_3$

3. When predicting a product for the reaction between two ionic compounds, we can always eliminate some of the ion pairs as possible products. Give a reason for eliminating each of the pairs as a product of the reaction below.

   $AgNO_3 + Na_3PO_4 \rightarrow$

   a. Ag^+, Na^+

   b. Na^+, NO_3^-

   c. NO_3^-, PO_3^{3-}

4. Use the solubility rules to predict the water solubility of each of the following compounds.

   a. Na_2S

   b. $PbCl_2$

   c. K_2SO_4

   d. $(NH_4)_2CrO_4$

   e. $Pb(OH)_2$

   f. $Ca(NO_3)_2$

   g. $Ba_3(PO_4)_2$

   h. $ZnCl_2$

5. For each word description, write the balanced molecular equation, and identify the product of the reaction.

   a. Aqueous solutions of sodium sulfate and lead(II) nitrate are mixed. One of the products is a white solid.

   b. Aqueous solutions of potassium hydroxide and nickel(II) chloride are mixed. One of the products is a green solid.

   c. Aqueous solutions of potassium sulfide and zinc nitrate are mixed. A pale yellow solid is produced.

   d. Aqueous solutions of silver nitrate and ammonium phosphate are mixed. A white solid is produced.

6. For each of the balanced equations below, write the complete ionic equation.

   a. $3CaCl_2(aq) + 2Na_3PO_4(aq) \rightarrow Ca_3(PO_4)_2(s) + 6NaCl(aq)$

   b. $Cu(NO_3)_2(aq) + K_2S(aq) \rightarrow CuS(s) + 2KNO_3(aq)$

   c. $2AgNO_3(aq) + K_2SO_4(aq) \rightarrow Ag_2SO_4(s) + 2KNO_3(aq)$

7. Complete and balance the equations below and identify the spectator ions.

   a. $Ca(NO_3)_2(aq) + K_2SO_4(aq) \rightarrow$

   b. $(NH_4)_2CO_3(aq) + CuCl_2(aq) \rightarrow$

   c. $NaOH(aq) + Pb(NO_3)_2(aq) \rightarrow$

   d. $Na_2S(aq) + Zn(NO_3)_2(aq) \rightarrow$

   e. $CoCl_2(aq) + Ca(OH)_2(aq) \rightarrow$

8. Write the net ionic equation for each reaction.

   a. $K_2CO_3(aq) + CaCl_2(aq) \rightarrow CaCO_3(s) + 2KCl(aq)$

   b. $Pb(NO_3)_2(aq) + (NH_4)_2S(aq) \rightarrow 2NH_4NO_3(aq) + PbS(s)$

   c. $2LiCl(aq) + 2Hg_2(NO_3)_2(aq) \rightarrow Hg_2Cl_2(s) + 2LiNO_3(aq)$

   d. $2NaOH(aq) + MgCl_2(aq) \rightarrow Mg(OH)_2(s) + 2NaCl(aq)$

9. What salts in aqueous solutions could you mix together to produce the solids below?

   a. $Zn(OH)_2$

   b. $Ba_3(PO_4)_2$

   c. $PbCl_2$

   d. $CaSO_4$

   e. $CoCO_3$

   f. Ag_2SO_4

10. Which of the substances below are strong acids, which are strong bases, and which are neither of these?

    a. HNO_3

    b. $C_2H_4O_2$

    c. H_2SO_4

    d. HCl

    e. NaCl

    f. K_2SO_4

11. Write complete ionic equations for the reactions below.

    a. Sodium hydroxide reacts with sulfuric acid.

    b. Hydrochloric acid reacts with potassium hydroxide.

    c. Nitric acid reacts with sodium hydroxide.

12. Write net ionic equations for each reaction in Problem 11.

13. Which of the reactions below are acid/base reactions?

    a. $K_2SO_4(aq) + Pb(NO_3)(aq) \rightarrow PbSO_4(s) + 2KNO_3(aq)$

    b. $KOH(aq) + HNO_3(aq) \rightarrow KNO_3(aq) + H_2O(l)$

    c. $H_2SO_4(aq) + 2NaOH(aq) \rightarrow Na_2SO_4(aq) + 2H_2O(l)$

    d. $Na_2CO_3(aq) + CoCl_2(aq) \rightarrow CoCO_3(s) + 2NaCl(aq)$

14. How many electrons do the elements below either gain or lose? For example, potassium atoms lose one electron.

$$K \rightarrow K^+ + e^-$$

a. Br_2

b. Mg

c. H_2

d. Al

e. O_2

f. S

15. Show how the ions below can gain or lose electrons to form atoms or molecules. For example, a sodium ion gains one electron to form an atom of sodium.

$$Na^+ + e^- \rightarrow Na$$

a. $2Cl^-$

b. K^+

c. $4P^{3+}$

d. Ca^{2+}

e. $2I^-$

f. Al^{3+}

16. For each reaction below, write equations showing the gain and loss of electrons.

a. $Cu(s) + 2AgNO_3(aq) \rightarrow 2Ag(s) + Cu(NO_3)_2(aq)$

b. $2HCl(aq) + Zn(s) \rightarrow H_2(g) + ZnCl_2(aq)$

c. $2NaBr(aq) + Cl_2(g) \rightarrow 2NaCl(aq) + Br_2(g)$

d. $2Hg(l) + O_2(g) \rightarrow 2HgO(s)$

17. Classify the reactions below as a precipitation reaction, an acid–base reaction, or an oxidation–reduction reaction.

a. $2NaCl(s) + Br_2(l) \rightarrow 2NaBr(s) + Cl_2(g)$

b. $Na_2SO_4(aq) + Pb(NO_3)_2(aq) \rightarrow PbSO_4(s) + 2NaNO_3(aq)$

c. $2NaOH(aq) + H_2SO_4(aq) \rightarrow 2H_2O(l) + Na_2SO_4(aq)$

d. $2AgNO_3(aq) + Fe(s) \rightarrow Fe(NO_3)_2(aq) + 2Ag(s)$

e. $2KOH(aq) + ZnCl_2(aq) \rightarrow Zn(OH)_2(s) + 2KCl(aq)$

18. Classify the reactions below as combustion, synthesis or decomposition reactions.

a. $N_2(g) + 3H_2(g) \rightarrow 2NH_3(g)$

b. $C_7H_{16}(g) + 11O_2(g) \rightarrow 7CO_2(g) + 8H_2O(g)$

c. $16Cu(s) + S_8(s) \rightarrow 8Cu_2S(s)$

d. $2NaNO_3(s) \rightarrow 2NaNO_2(s) + O_2(g)$

e. $SO_3(g) + H_2O(l) \rightarrow H_2SO_4(l)$

19. Write balanced equations for each of the word descriptions. Classify each reaction as precipitation, oxidation–reduction, or acid–base.

    a. Ethyl alcohol, a gasoline additive, burns in the presence of oxygen gas to produce carbon dioxide and water vapor.

    b. Aqueous solutions of ammonium sulfide and lead nitrate are mixed to produce solid lead sulfide and aqueous ammonium nitrate.

    c. Aluminum metal reacts with oxygen to produce solid aluminum oxide.

    d. Sodium metal reacts with liquid water to produce aqueous sodium hydroxide and hydrogen gas.

    e. Aqueous solutions of potassium hydroxide and nitric acid are mixed to produce aqueous potassium nitrate and liquid water.

    f. Aqueous solutions of sodium phosphate and aqueous silver nitrate are mixed to produce solid silver phosphate and aqueous sodium nitrate.

ANSWERS TO LEARNING REVIEW

1. Only "c," color change, does not tend to make a reaction occur.

2.
    a. $2NH_4^+$ $1SO_4^{2-}$
    b. $1K^+$ $1NO_3^-$
    c. $2Na^+$ $1Cr_2O_7^{2-}$
    d. $1Mg^{2+}$ $2Cl^-$
    e. $3Li^+$ $1PO_4^{3-}$
    f. $1Al^{3+}$ $3NO_3^-$

3.
    a. Both Ag^+ and Na^+ are cations. An anion and a cation are needed to form a neutral product.
    b. $NaNO_3$ is soluble in water, and so it exists in solution as Na^+ ions and NO_3^- ions.
    c. Both NO_3^- and PO_4^{3-} are anions. An anion and a cation are needed to form a neutral product.

4.
    a. water soluble (Rule 2)
    b. not soluble (Rule 3)
    c. water soluble (Rule 2)
    d. water soluble (Rule 2)
    e. not soluble (Rule 5)
    f. water soluble (Rule 1)
    g. not soluble (Rule 6)
    h. water soluble (Rule 3)

5.

a. $Na_2SO_4(aq) + Pb(NO_3)_2(aq) \rightarrow PbSO_4(s) + 2NaNO_3(aq)$

The product is lead(II) sulfate.

b. $NiCl_2(aq) + 2KOH(aq) \rightarrow Ni(OH)_2(s) + 2KCl(aq)$

The product is nickel(II) hydroxide.

c. $K_2S(aq) + Zn(NO_3)_2(aq) \rightarrow ZnS(s) + 2KNO_3(aq)$

The product is zinc sulfide.

d. $3AgNO_3(aq) + (NH_4)_3PO_4(aq) \rightarrow Ag_3PO_4(s) + 3NH_4NO_3(aq)$

The product is silver phosphate.

6.

a. $3Ca^{2+} + 6Cl^- + 6Na^+ + 2PO_4^{3-} \rightarrow Ca_3(PO_4)_2(s) + 6Na^+ + 6Cl^-$

b. $Cu^{2+} + 2NO_3^- + 2K^+ + S^{2-} \rightarrow CuS(s) + 2K^+ + 2NO_3^-$

c. $2Ag^+ + 2NO_3^- + 2K^+ + SO_4^{2-} \rightarrow Ag_2SO_4(s) + 2K^+ + 2NO_3^-$

7.

a. $Ca(NO_3)_2(aq) + K_2SO_4(aq) \rightarrow CaSO_4(s) + 2KNO_3(aq)$

K^+ and NO_3^- are the spectator ions.

b. $(NH_4)_2CO_3(aq) + CuCl_2(aq) \rightarrow CuCO_3(s) + 2NH_4Cl(aq)$

NH_4^+ and Cl^- are the spectator ions.

c. $2NaOH(aq) + Pb(NO_3)_2(aq) \rightarrow Pb(OH)_2(s) + 2NaNO_3(aq)$

Na^+ and NO_3^- are the spectator ions.

d. $Na_2S(aq) + Zn(NO_3)_2(aq) \rightarrow ZnS(s) + 2NaNO_3(aq)$

Na^+ and NO_3^- are the spectator ions.

e. $CoCl_2(aq) + Ca(OH)_2(aq) \rightarrow Co(OH)_2(s) + CaCl_2(aq)$

Ca^{2+} and Cl^- are the spectator ions.

8.

a. $Ca^{2+} + CO_3^{2-} \rightarrow CaCO_3(s)$

b. $Pb^{2+} + S^{2-} \rightarrow PbS(s)$

c. $Hg_2^{2+} + 2Cl^- \rightarrow Hg_2Cl_2(s)$

d. $Mg^{2+} + 2OH^- \rightarrow Mg(OH)_2(s)$

9. It is possible to produce the solids below from several different soluble salts, so your answer could be correct and not the same as the answer below. If your answer does not match the one below, use the solubility rules to help you determine whether the aqueous salt solutions you chose would be soluble in water and whether an exchange of anions would produce the desired insoluble salt.

a. $Zn(NO_3)_2$ and $NaOH$

b. $Ba(NO_3)_2$ and K_3PO_4

c. $Pb(NO_3)_2$ and NaCl

d. $CaCl_2$ and $(NH_4)_2SO_4$

e. $Co(NO_3)_2$ and Na_2CO_3

f. $AgNO_3$ and K_2SO_4

10.

a. HNO_3 is a strong acid.

b. $C_2H_4O_2$ is a weak acid, so the correct answer is neither of these.

c. H_2SO_4 is a strong acid.

d. HCl is a strong acid.

e. NaCl is a salt produced when HCl and NaOH react, so it is neither a strong acid nor a strong base.

f. K_2SO_4 is a salt produced when H_2SO_4 and KOH react, so it is neither a strong acid nor a strong base.

11.

a. $2Na^+ + 2OH^- + 2H^+ + SO_4^{2-} \rightarrow 2Na^+ + SO_4^{2-} + 2H_2O(l)$

b. $K^+ + OH^- + H^+ + Cl^- \rightarrow K^+ + Cl^- + H_2O(l)$

c. $Na^+ + OH^- + H^+ + NO_3^- \rightarrow Na^+ + NO_3^- + H_2O(l)$

12.

a. $H^+ + OH^- \rightarrow H_2O(l)$

b. $H^+ + OH^- \rightarrow H_2O(l)$

c. $H^+ + OH^- \rightarrow H_2O(l)$

13.

a. This is a precipitation reaction.

b. This is an acid–base reaction. The products are water and the salt KNO_3.

c. This is an acid–base reaction. The products are water and the salt Na_2SO_4.

d. This is a precipitation reaction.

14. When atoms or molecules lose electrons, a positively charged cation is produced. When electrons are gained, then a negatively charged anion is produced. You can show how electrons are gained or lost by adding electrons to either the right side or the left side of an equation.

a. $Br_2 + 2e^- \rightarrow 2Br^-$

b. $Mg \rightarrow Mg^{2+} + 2e^-$

c. $H_2 \rightarrow 2H^+ + 2e^-$

d. $Al \rightarrow Al^{3+} + 3e^-$

e. $O_2 + 4e^- \rightarrow 2O^{2-}$

f. $S + 2e^- \rightarrow S^{2-}$

15. Ions can either gain or lose electrons to become neutral atoms or molecules. You can show whether the ions must lose or gain electrons by adding electrons to either the right side or the left side of an equation.

    a. $2Cl^- \rightarrow Cl_2 + 2e^-$

    b. $K^+ + e^- \rightarrow K$

    c. $4P^{3-} \rightarrow P_4 + 12e^-$

    d. $Ca^{2+} + 2e^- \rightarrow Ca$

    e. $2I^- \rightarrow I_2 + 2e^-$

    f. $Al^{3+} + 3e^- \rightarrow Al$

16. When presented with a reaction where electrons are transferred, it is possible to extract the parts of the reaction where electrons are lost and where electrons are gained and to write each part separately. Notice that the number of electrons lost is equal to the number gained.

    a. $Cu \rightarrow Cu^{2+} + 2e^-$ Two electrons are lost
    $2Ag^+ + 2e^- \rightarrow 2Ag$ Two electrons are gained

    b. $Zn \rightarrow Zn^{2+} + 2e^-$ Two electrons are lost
    $2H^+ + 2e^- \rightarrow H_2$ Two electrons are gained

    c. $2Br^- \rightarrow Br_2 + 2e^-$ Two electrons are lost
    $Cl_2 + 2e^- \rightarrow H_2$ Two electrons are gained

    d. $2Hg \rightarrow 2Hg^{2+} + 4e^-$ Four electrons are lost
    $O_2 + 4e^- \rightarrow 2O^{2-}$ Four electrons are gained

17.

    a. In this reaction, two chloride ions lose electrons to become a chlorine molecule, and a bromine molecule gains two electrons to become two bromide ions. This is an oxidation–reduction reaction. Because the chloride ion paired with sodium is exchanged for a bromide ion, this kind of reaction is often called a replacement reaction.

    b. Two aqueous solutions containing ionic compounds react, and one of the products is the ionic solid $PbSO_4$. This is a precipitation reaction.

    c. The base NaOH reacts with the acid H_2SO_4 to produce water. This is an acid–base reaction.

    d. In this reaction, two Ag^+ ions gain two electrons to become two atoms of silver, and an atom of iron loses two electrons to become an Fe^{2+} ion. This is an oxidation–reduction reaction. Because the silver ion paired with the nitrate ion is exchanged for an iron ion, this kind of reaction is called a replacement reaction.

    e. Two aqueous solutions containing ionic compounds react and one of the products is the ionic solid $Zn(OH)_2$. This is an example of a precipitation reaction.

18.

a. Molecular nitrogen and molecular hydrogen react to produce a larger molecule, ammonia. This is a synthesis reaction.

b. A molecule that is composed of carbon and hydrogen reacts with oxygen gas. The products are carbon dioxide and water. Reactions that have oxygen as a reactant are members of a sub-class of oxidation–reduction reactions called combustion reactions.

c. Elemental copper reacts with elemental sulfur. A compound containing both elements is the product. This is a synthesis reaction.

d. Solid sodium nitrate is converted to two simpler molecules, sodium nitrite and molecular oxygen. This is an example of a decomposition reaction.

e. In this reaction, two small molecules combine to produce one larger molecule. This is an example of a synthesis reaction.

19.

a. $C_2H_5OH(l) + 3O_2(g) \rightarrow 2CO_2(g) + 3H_2O(l)$

A molecule reacts with oxygen gas. Because this reaction has oxygen as a reactant, it is an oxidation–reduction reaction.

b. $(NH_4)_2S(aq) + Pb(NO_3)_2(aq) \rightarrow PbS(s) + 2NH_4NO_3(aq)$

Two aqueous solutions are mixed to produce a solid product (precipitation reaction).

c. $4Al(s) + 3O_2(g) \rightarrow 2Al_2O_3(s)$

Elemental aluminum loses three electrons to become Al^{3+}, and molecular oxygen gains two electrons to become O^{2-}. This is an oxidation–reduction reaction.

d. $2Na(s) + 2H_2O(l) \rightarrow 2NaOH(aq) + H_2(g)$

Sodium metal loses an electron to become Na^+, and two hydrogen ions gain an electron to become hydrogen gas. This is an oxidation–reduction reaction.

e. $KOH(aq) + HNO_3(aq) \rightarrow KNO_3(aq) + H_2O(l)$

Aqueous solutions of the base KOH and the acid HNO_3 are mixed to produce liquid water, so this is an acid–base reaction.

f. $Na_3PO_4(aq) + 3AgNO_3(aq) \rightarrow Ag_3PO_4(s) + 3NaNO_3(aq)$

Two aqueous solutions are mixed. The product is a solid, Ag_3PO_4, so this is a precipitation reaction.

CHAPTER 8

Chemical Composition

INTRODUCTION

Before beginning a project of any kind, it is always important to know the quantity of material needed to finish the project. It is usually possible to count the number of individual items you will need. In chemistry, it is difficult to count the number of atoms or molecules needed because the individual particles are too small, and there are too many of them. This chapter will show you how you can count the number of particles by weighing them.

CHAPTER DISCUSSION

The Mole

One crucial concept in this chapter is that of the mole. Make sure you understand why it is so important. There are two main ideas you need to consider:

1. We can count objects by weighing a sample of the objects provided we know the average mass of the objects.
2. Relative masses of two or more different objects stay the same but can be in larger units if we have the same number of objects.

Section 8.1 in your text provides a very good discussion of this first point. Make sure to read this. Talk to an instructor if you have difficulty with it. Let's look at the second point more carefully.

Suppose we have two blocks, a red block and a yellow block. The red block weighs 1.0 ounce, and the yellow block weighs 4.0 ounces. Now suppose we have 16 of each block. What is the mass of each sample? The sample of red blocks weighs 16.0 ounces, and the sample of yellow blocks weighs 64.0 ounces. But note that 16.0 ounces is also 1.0 pound, thus 64.0 ounces is 4.0 pounds. Note the relative masses of the blocks:

	One Block	Sixteen Blocks
red	1.0 ounce	1.0 pound
yellow	4.0 ounces	4.0 pounds

The relative masses stay the same (1:4) but the units are changed. Why is this important?

Recall from Chapter 4 that the periodic table gives us the relative masses of the elements. But what are the units of these? Actually, there need not be any unit at all, the units could be anything; that is, the table gives us relative masses much like the 1:4 ratio we see in the examples with the blocks. But what units would be useful for us? The standard unit of mass that we will use is the gram. However, the average hydrogen atom, for example, has a mass of 1.66057×10^{-24} g. This is much too small for us to deal with. We would like to keep the relative mass of hydrogen at 1.008 (as it is on the periodic table) but in units of grams. How many hydrogen atoms are there in a 1.008-g sample of hydrogen?

$$1.008 \text{ g H} \times \left(\frac{1 \text{ atom H}}{1.66057 \times 10^{-24} \text{ g H}}\right) = 6.022 \times 10^{23} \text{ H atoms}$$

Thus, if we have 6.022×10^{23} atoms of hydrogen the sample will have a mass of 1.008 g. Along the same lines, if we have 6.022×10^{23} atoms of carbon, the sample will have a mass of 12.01 g. If we have 6.022×10^{23} atoms of oxygen, the sample will have a mass of 16.00 g. That is, 6.022×10^{23}, which is called a mole, is the number that converts the units on the periodic table (called amu or atomic mass units) to grams.

The mole is just a number (like a dozen is 12), and it allows us to convert between how many atoms (or molecules) we have and the mass of the sample. In the example with the blocks, this number was 16 (to convert ounces to grams). To convert amu to grams, we use the mole.

Formulas and Mass Percent

Would you say ammonia (NH_3) is mostly nitrogen or mostly hydrogen? Your answer depends on if you are looking at the number of atoms or the mass of the atoms. In terms of numbers of atoms, ammonia is ¾ hydrogen (there are four atoms making up an ammonia molecule, and three of them are hydrogen). But it is often important to know the composition by mass of a compound. Chemists have instruments that give them percent by mass data, and they use this to determine the formulas of compounds. How can we do this?

Suppose we have 1.0 mole of ammonia molecules. The molar mass is 17.034 g (N = 14.01 g/mol, and each hydrogen is 1.008 g/mol, so the molar mass of NH_3 is 14.01 + 3(1.008) = 17.034 g). How much of this mass is hydrogen? How much of this mass is nitrogen?

Since there are three hydrogen atoms per ammonia molecule, there are three moles of hydrogen atoms per mole of ammonia molecule. Thus, the total mass of hydrogen should be 3(1.008) or 3.024 g, and there is 14.01 g of nitrogen. The mass percent of each element is:

$$\text{Hydrogen: } \frac{3.024 \text{ g}}{17.034 \text{ g}} \times 100\% = 17.75\% \text{ hydrogen by mass}$$

$$\text{Nitrogen: } \frac{14.01 \text{ g}}{17.034 \text{ g}} \times 100\% = 82.25\% \text{ nitrogen by mass}$$

Note that even though there are more hydrogen atoms than nitrogen atoms, the percent by mass of hydrogen in this case is lower than that of nitrogen.

Try the following example before reading on:

What is the mass percent of hydrogen and of nitrogen for the compound N_2H_6?

To solve this, you can find the molar mass of the compound, which is 34.068, and the total mass of hydrogen (6.048 g) and nitrogen (28.02 g). If you are having difficulty getting these numbers, read through your text or talk with your instructor.

Thus, the mass percent of each element is:

$$\text{Hydrogen: } \frac{6.028 \text{ g}}{34.068 \text{ g}} \times 100\% = 17.75\% \text{ hydrogen by mass}$$

$$\text{Nitrogen: } \frac{28.02 \text{ g}}{34.068 \text{ g}} \times 100\% = 82.25\% \text{ nitrogen by mass}$$

Why is this significant? Note that this is the same percent by mass as ammonia (NH_3). Chemists generally use percent-by-mass data to determine the formula of a compound. But what if we know a compound is 17.75% hydrogen by mass and 82.25% nitrogen by mass? Is the formula NH_3? Or N_2H_6? Actually, you should prove to yourself that any formula that has 3 times as many hydrogen atoms as nitrogen atoms (N_xH_{3x}) will be 17.75% hydrogen and 82.25% nitrogen by mass. So what are we to do?

This is why we need to know the molar mass of a compound to know the actual formula for the compound (the molecular formula). Given just the percent-by-mass data allows us to determine only the ratio of atoms in the molecule (in this case 1:3) and determine the empirical formula. Let's consider an example.

> A compound consisting of carbon, hydrogen, and oxygen is 40.00% carbon by mass and 6.71% hydrogen by mass. What is the empirical formula of the compound?

In this case, we are given the percent-by-mass data and are asked to determine the formula for the compound. How should we think about this?

We are looking for the formula (empirical) of this compound. Recall that the formula gives us the ratio of atoms in the compound. The general formula for this compound is

$$C_xH_yO_z$$

where x, y, and z represent the number of atoms in one molecule of the compound (or, in the case of the empirical formula, the lowest whole-number ratio). Our problem, then, is to determine these numbers. Thus, we have been given mass-percent data for the atoms and need to determine the number of atoms. In essence, we will have to change a mass to a number. How do we do this? Well, we know, for example, that 1 mole (a number, 6.022×10^{23}) of carbon atoms has a mass of 12.01 g. Thus, we will use the molar mass of the atoms to make this conversion.

Before we go on, notice that we have done quite a bit of thinking about this problem before doing any calculations. This is generally a good approach with chemistry problems. That is, think about the setup of the problem before worrying about the specifics. Do not immediately try to plug numbers into a given equation. Think about the problem first.

Now we know that we will need to convert the mass data to numbers using the mole. But what is the mass of carbon? Hydrogen? Oxygen? We are only given the mass-percent data. This requires us to think about what is meant by mass percent.

Stating that the compound is 40.00% carbon by mass means that for every 100.00 g of compound, 40.00 g is carbon.

It is not stating that we have 40.00 g of carbon necessarily, but that 40.00 g out of every 100.00 g of compound is carbon (if we had 200.00 g of compound, we would have 80.00 g of carbon, for example). Since we are trying to determine the number ratio for the atoms, we need to know only the mass ratio, not the actual masses. The easiest way to do this is to assume we have 100.00 g of the compound (we could assume <u>any</u> mass here; make sure you understand why). Therefore, the masses of the atoms are:

mass of carbon:	40.00 g
mass of hydrogen:	6.71 g
mass of oxygen:	53.29 g

Note that these masses add up to 100.00 g (which is how we can determine the value for oxygen). Now we can convert these masses to numbers (moles) by using the molar masses of each element. Do this before reading on.

The mole ratios you should have calculated are

moles of carbon:	3.33
moles of hydrogen:	6.66
moles of oxygen:	3.33

If you cannot get these numbers, review this in your text, work with a friend, or see your instructor.

We now have the mole ratios. Does this mean the empirical formula is $C_{3.33}H_{6.66}O_{3.33}$? No. We must represent the formulas with whole numbers (we cannot have fractions of atoms). Note, however, that 3.33:6.66:3.33 can be written as 1:2:1. Since we used mass-ratio data, we calculated atom-ratio data. That is, we do not actually have 3.33 moles of carbon (or 2 moles for that matter) but we have a 1:2:1 ratio of atoms of carbon, hydrogen, and oxygen, respectively. Thus, we can write the empirical formula as

$$C_1H_2O_1$$

But this is not necessarily the actual (molecular) formula of the compound. You should prove to yourself, for example, that the formula $C_2H_4O_2$ has the same mass percent of each atom as $C_1H_2O_1$. To determine the actual formula we need to know the molar mass of the compound.

For example, if we are told that the molar mass of this compound is about 180 g/mol, how do we determine the molecular formula? We know that the answer has to have the general formula $C_xH_{2x}O_x$. Recall that the molar mass is the sum of the masses of the atoms. Thus, $180 = 12.01(x) + 1.008(2x) + 16.00(x)$, or $180 = 30.026x$. Solving for x, we get 6. Therefore, the molecular formula is $C_6H_{12}O_6$.

Again, note the amount of thought that went into solving this problem. Do not expect to simply plug numbers into equations. By thinking about the underlying concepts involved, you will find that you can solve quite difficult and novel problems. This is a major goal.

LEARNING REVIEW

1. A hardware store employee determined that the average mass of a certain size nail was 2.35 g.
   a. How many nails are there in 1057.5 g nails?
   b. If a customer needs 1500 nails, what mass of nails should the employee weigh out?
2. Ten individual screws have masses of 10.23 g, 10.19 g, 10.24 g, 10.23 g, 10.26 g, 10.23 g, 10.28 g, 10.30 g, 10.25 g, and 10.26 g. What is the average mass of a screw?
3. The average mass of a hydrogen atom is 1.008 amu. How many hydrogen atoms are there in a sample that has a mass of 25,527.6 amu?
4. The average mass of a sodium atom is 22.99 amu. What is the mass in amu of a sample of sodium atoms that contains 3.29×10^3 sodium atoms?
5. A sample with a mass of 4.100×10^5 amu is 25.00% carbon, and 75.00% hydrogen by mass. How many atoms of carbon and hydrogen are in the sample? The average mass of a carbon atom is 12.01 amu and of a hydrogen atom is 1.008 amu.
6. The average mass of a neon atom is 20.18 amu.
   a. How many grams of neon are found in a mole of neon?
   b. How many atoms of neon are in a mole of neon?
7. What is the value of Avogadro's number, and how is it defined?

8. Use the average mass values found inside the front cover of your textbook to solve the problems below.

   a. A helium balloon contains 5.38×10^{22} helium atoms. How many grams of helium are in the balloon?

   b. A piece of iron was found to contain 3.25 mol Fe. How many grams are in the sample?

   c. A sample of liquid bromine contains 65.00 g Br atoms. How many bromine atoms are in the sample?

   d. A sample of zinc contains 0.78 mol Zn. How many zinc atoms are in this sample?

9. What is meant by the term "molar mass"?

10. Calculate the molar mass of the following substances.

    a. Fe_2O_3

    b. NH_3

    c. C_2H_5OH

    d. CO_2

    e. N_2O_5

11. Calculate the molar mass of these ionic compounds.

    a. HCl

    b. $MgBr_2$

    c. $Pb(OH)_2$

    d. $Cu(NO_3)_2$

    e. KCl

    f. Na_2SO_4

12. Acetone, which has a formula of C_3H_6O, is used as a solvent in some fingernail polish removers. How many moles of acetone are in 5.00 g of acetone?

13. How many grams of potassium sulfate are in 0.623 mol potassium sulfate?

14. Calculate the mass fraction of nitrogen in N_2O_5.

15. Calculate the mass percent of each element in the following substances.

    a. CH_3NH_2

    b. H_2SO_4

16. Explain the difference between the empirical formula and the molecular formula of a compound.

17. The molecular formula of the gas, acetylene, is C_2H_2. What is the empirical formula?

18. When 2.500 g of an oxide of mercury, Hg_xO_y, is decomposed into the elements by heating, 2.405 g of mercury is produced. Calculate the empirical formula for this compound.

Hg_xO_y	$\rightarrow$	x Hg	+	y O
2.500 g		2.405 g		? g

19. A compound was analyzed and found to contain only carbon, hydrogen and chlorine. A 6.380-g sample of the compound contained 2.927 g carbon and 0.5729 g hydrogen. What is the empirical formula of the compound?

20. The compound benzamide has the following percent composition. What is the empirical formula?

$$C = 69.40\% \quad H = 5.825\% \quad N = 11.57\% \quad O = 13.21\%$$

21. The empirical formula for a compound used in the past as a green paint pigment is $C_2H_3As_3Cu_2O_8$. The molar mass is 1013.71 g. What is the molecular formula?

22. A sugar that is broken down by the body to produce energy has the following percent composition.

$$C = 39.99\% \quad H = 6.713\% \quad O = 53.29\%$$

The molar mass is 210.18 g. What is the molecular formula?

ANSWERS TO THE LEARNING REVIEW

1.

a. This problem relies on the principle of counting by weighing. The question "how many nails?" can be answered because we are given the average mass of 1 nail.

$$\frac{1 \text{ nail}}{2.35 \text{ g}} \times 1057.5 \text{ g} = 450. \text{ nails}$$

b. If we know the mass of 1 nail equals 2.35 g, then the mass of 1500 nails is a multiple of 2.35 g.

$$\frac{2.35 \text{ g}}{1 \text{ nail}} \times 1500. \text{ nails} = 3530 \text{ g}$$

2. Average mass can be determined by adding the masses of each individual screw, then dividing by the number of screws measured.

$$10.23 \text{ g} + 10.19 \text{ g} + 10.24 \text{ g} + 10.23 \text{ g} + 10.26 \text{ g} + 10.23 \text{ g} + 10.28 \text{ g} + 10.30 \text{ g} + 10.25 \text{ g} + 10.26 \text{ g} = 102.47 \text{ g}$$

The total mass of all 10 screws is 102.47 g.

$$\frac{102.47 \text{ g}}{10 \text{ screws}} = 10.25 \text{ g/screw}$$

The average mass of a screw is 10.25 g.

3. This problem is an example of counting by weighing. We are given the average mass of one hydrogen atom, and asked for the number of hydrogen atoms in some other mass of hydrogen.

$$\frac{1 \text{ hydrogen atom}}{1.008 \text{ amu}} \times 25{,}527.6 \text{ amu} = 25{,}330 \text{ hydrogen atoms}$$

4. If we know the average mass of an atom, we can calculate the mass of any quantity of atoms.

$$\frac{22.99 \text{ amu}}{1 \text{ sodium atom}} \times 3.29 \times 10^3 \text{ sodium atoms} = 75{,}600 \text{ amu}$$

5. The total mass of the sample is 4.100×10^5 amu. Of this mass, 25.00% comes from carbon atoms. So the mass contributed by carbon is

$$4.100 \times 10^5 \text{ amu} \times 0.2500 = 1.025 \times 10^5 \text{ amu}$$

The mass contributed by hydrogen is the original mass minus the mass contributed by carbon

$$4.100 \times 10^5 \text{ amu} - 1.025 \times 10^5 \text{ amu} = 3.075 \times 10^5 \text{ amu}$$

Now that we know the total mass of each kind of atom, we can use the average mass of one atom to count the number of atoms present.

$$\frac{1 \text{ hydrogen atom}}{1.008 \text{ amu}} \times 3.075 \times 10^5 \text{ amu} = 3.051 \times 10^5 \text{ hydrogen atoms}$$

$$\frac{1 \text{ carbon atom}}{12.01 \text{ amu}} \times 1.025 \times 10^5 \text{ amu} = 8.535 \times 10^3 \text{ carbon atoms}$$

6.

a. A mole of any element always contains a mass in grams equal to the average atomic mass of that element. So there are 20.18 g Ne in 1 mol Ne.

b. A mole of atoms of any element always contains 6.022×10^{23} atoms.

7. Avogadro's number is the number equal to the number of atoms in 12.01 grams of carbon. Chemists have accurately determined this number to be 6.022×10^{23} atoms.

8. These problems use conversions between moles and grams or between moles and number of atoms. For each element, you must write a different conversion factor for moles to grams depending upon the average mass for that element.

a. This problem requires first determining the moles of He and then converting moles to grams.

$$5.38 \times 10^{22} \text{ He atoms} \times \frac{1 \text{ mol He}}{6.022 \times 10^{23} \text{ He atoms}} \times \frac{4.003 \text{ g He}}{1 \text{ mol He}} = 0.357 \text{ g He}$$

b. $3.25 \text{ mol Fe} \times \frac{55.85 \text{ g Fe}}{1 \text{ mol Fe}} = 182 \text{ g Fe}$

c. This is a two-step problem requiring that you first calculate the number of moles of Br, then the number of Br atoms.

$$65.00 \text{ g Br} \times \frac{1 \text{ mol Br}}{79.90 \text{ g Br}} \times \frac{6.022 \times 10^{23} \text{ Br atoms}}{1 \text{ mol Br}} = 4.899 \times 10^{23} \text{ Br atoms}$$

d. $0.78 \text{ mol Zn} \times \frac{6.022 \times 10^{23} \text{ Zn atoms}}{1 \text{ mol Zn}} = 4.7 \times 10^{23} \text{ Zn atoms}$

9. Molar mass is the number of grams found in 1 mole of a substance. The molar mass is calculated by adding together the masses of each atom in the substance.

10.

a. Fe_2O_3 contains two Fe atoms and three O atoms.

$(2 \times 55.85 \text{ g Fe}) + (3 \times 16.00 \text{ g O}) = 159.7 \text{ g}$

b. NH_3 contains one nitrogen atom and three hydrogen atoms.

$(1 \times 14.01 \text{ g N}) + (3 \times 1.008 \text{ g H}) = 17.03 \text{ g}$

c. C_2H_5OH contains two C atoms, six H atoms and one O atom.

$(2 \times 12.01 \text{ g C}) + (6 \times 1.008 \text{ g H}) + (1 \times 16.00 \text{ g O}) = 46.07 \text{ g}$

d. CO_2 contains one C atom and two O atoms.

$(1 \times 12.01 \text{ g C}) + (2 \times 16.00 \text{ g O}) = 44.01 \text{ g}$

e. N_2O_5 contains two N atoms and five O atoms

$(2 \times 14.02 \text{ g N}) + (5 \times 16.00 \text{ g O}) = 108.0 \text{ g}$

11.

a. $1.008 \text{ g H} + 35.45 \text{ g Cl} = 36.46 \text{ g}$

b. $24.31 \text{ g Mg} + (2 \times 79.90 \text{ g Br}) = 184.1 \text{ g}$

c. $207.19 \text{ g Pb} + (2 \times 16.00 \text{ g O}) + (2 \times 1.008 \text{ H}) = 241.2 \text{ g}$

d. $63.55 \text{ g Cu} + (2 \times 14.01 \text{ g N}) + (6 \times 16.00 \text{ g O}) = 187.6 \text{ g}$

e. $39.10 \text{ g K} + 35.45 \text{ g Cl} = 74.55 \text{ g}$

f. $(2 \times 22.99 \text{ g Na}) + 32.07 \text{ g S} + (4 \times 16 \text{ g O}) = 142.1 \text{ g}$

12. To solve this problem, we need to know how many grams of acetone are in one mole of acetone. The number of grams of acetone equal to one mole of acetone is the molar mass.

molar mass acetone $= (3 \times 12.01 \text{ g}) + (6 \times 1.008 \text{ g}) + 16.00 \text{ g} = 58.08 \text{ g}$

$$5.00 \text{ g acetone} \times \frac{1 \text{ mol acetone}}{58.08 \text{ g acetone}} = 0.0861 \text{ mol acetone}$$

13. molar mass $K_2SO_4 = (2 \times 39.10 \text{ g}) + 32.07 \text{ g} + (4 \times 16.00 \text{ g}) = 174.3 \text{ g}$

$$0.623 \text{ mol } K_2SO_4 \times \frac{174.3 \text{ g } K_2SO_4}{1 \text{ mol } K_2SO_4} = 109 \text{ g } K_2SO_4$$

14. Mass fraction is equal to the mass of the desired element, in this case nitrogen, divided by the molar mass.

$$\frac{28.02 \text{ g N}}{108.0 \text{ g total}} = 0.2594$$

15.

a. molar mass of CH_3NH_2 is $12.01 \text{ g C} + (5 \times 1.008 \text{ g H}) + 14.01 \text{ g N} = 31.06 \text{ g total}$

$$\frac{12.01 \text{ g C}}{31.06 \text{ g total}} \times 100 = 38.67\% \text{ C}$$

$$\frac{5.040 \text{ g H}}{31.06 \text{ g total}} \times 100 = 16.23\% \text{ H}$$

$$\frac{14.01 \text{ g N}}{31.06 \text{ g total}} \times 100 = 45.11\% \text{ N}$$

b. molar mass of H_2SO_4 is $(2 \times 1.008 \text{ g H}) + 32.07 \text{ g S} + (4 \times 16.00 \text{ g O}) = 98.09 \text{ g total}$

$$\frac{2.016 \text{ g H}}{98.09 \text{ g total}} \times 100 = 2.055\% \text{ H}$$

$$\frac{32.07 \text{ g S}}{98.09 \text{ g total}} \times 100 = 32.69\% \text{ S}$$

$$\frac{64.00 \text{ g O}}{98.09 \text{ g total}} \times 100 = 65.25\% \text{ O}$$

16. The empirical formula gives only the relative number of atoms, that is, a ratio of each kind of atom. The molecular formula tells exactly how many of each kind of atom are present in the molecule.

17. For every two atoms of carbon in acetylene there are two atoms of hydrogen. The ratio of carbon atoms to hydrogen atoms is 1:1. So the empirical formula of acetylene is CH.

18. Since we know that mercury and the oxygen combined weighed 2.500 g before the reaction took place and that the mass of the mercury is 2.405 g, then the mass of oxygen must be $2.500 - 2.405 = 0.095$ g. We can now convert grams of mercury and grams of oxygen to moles using the atomic masses of these elements.

$$2.405 \text{ g Hg} \times \frac{1 \text{ mol Hg}}{200.59 \text{ g Hg}} = 0.01199 \text{ mol Hg}$$

$$0.095 \text{ g O} \times \frac{1 \text{ mol O}}{16.00 \text{ g O}} = 0.0059 \text{ mol O}$$

The ratio of mercury atoms to oxygen atoms is, $\frac{0.01199}{0.0059} = 2.03$ to 1.

So, there are twice as many Hg atoms as O atoms, and the empirical formula is Hg_2O.

19. When 6.380 g of a compound that contained only carbon, hydrogen and chlorine was analyzed, it was found to contain 2.927 g carbon and 0.5729 g hydrogen. The mass of chlorine must be equal to the total mass minus the mass of carbon plus hydrogen.

$$\text{mass of chlorine} = 6.380 - (2.927 \text{ g C} + 0.5729 \text{ g H}) = 2.880 \text{ g}$$

The moles of each kind of atom are determined from the average atomic mass.

$$2.927 \text{ g C} \times \frac{1 \text{ mol C}}{12.01 \text{ g C}} = 0.2437 \text{ mol C}$$

$$0.5729 \text{ g H} \times \frac{1 \text{ mol H}}{1.008 \text{ g H}} = 0.5684 \text{ mol H}$$

$$2.880 \text{ g Cl} \times \frac{1 \text{ mol Cl}}{35.45 \text{ g Cl}} = 0.08124 \text{ mol Cl}$$

Express the mole ratios in whole numbers by dividing each number of moles by the smallest number of moles.

$$\frac{0.2437 \text{ mol C}}{0.08124} = 3.000 \text{ mol C}$$

$$\frac{0.5684 \text{ mol H}}{0.08124} = 7.000 \text{ mol H}$$

$$\frac{0.08124 \text{ mol Cl}}{0.08124} = 1.000 \text{ mol Cl}$$

The empirical formula is C_3H_7Cl.

20. This problem provides only percent composition data for the compound benzamide. It does not provide an analysis in grams for each of the elements present. We need to know how many grams of each element are present in a sample of benzamide so we can calculate the moles of each element. We can convert percent composition data to grams of each element. Assume that we have 100.0 g of benzamide. Of that sample, 69.40% is carbon. So for a 100.0 g sample, 69.40 g are carbon, 5.825 g are hydrogen, 11.57 g are nitrogen, and 13.21 g are oxygen. We can calculate the number of moles of each element.

$$69.40 \text{ g C} \times \frac{1 \text{ mol C}}{12.01 \text{ g C}} = 5.779 \text{ mol C}$$

$$5.825 \text{ g H} \times \frac{1 \text{ mol H}}{1.008 \text{ g H}} = 5.779 \text{ mol H}$$

$$11.57 \text{ g N} \times \frac{1 \text{ mol N}}{14.01 \text{ g N}} = 0.8258 \text{ mol N}$$

$$13.21 \text{ g O} \times \frac{1 \text{ mol O}}{16.00 \text{ g O}} = 0.8256 \text{ mol O}$$

Now divide each number of moles by the smallest number of moles to convert the number of moles to whole numbers.

$$\frac{5.779 \text{ mol C}}{0.08256} = 7.000 \text{ mol C}$$

$$\frac{5.779 \text{ mol H}}{0.08256} = 7.000 \text{ mol H}$$

$$\frac{0.08258 \text{ mol N}}{0.08256} = 1.000 \text{ mol N}$$

$$\frac{0.8256 \text{ mol O}}{0.08256} = 1.000 \text{ mol O}$$

The empirical formula is C_7H_7NO.

21. If you are given both the molar mass and the empirical formula, determining the molecular formula is straightforward. If we multiply all the atoms in the empirical formula by some number, we will have a correct molecular formula. So the molecular formula is a multiple of the empirical formula. We can determine what this multiple is by comparing the molar mass of the molecular formula with the molar mass of the empirical formula.

$$\text{molar mass empirical formula} = (2 \times 12.01 \text{ g C}) + (3 \times 1.008 \text{ g H}) + (3 \times 74.92 \text{ g As}) + (2 \times 63.55 \text{ g Cu}) + (8 \times 16.00 \text{ g O}) = 506.9 \text{ g}$$

The molar mass of the empirical formula is 506.9 g, and we know that the molar mass of the molecular formula is 1013.7 g. There are two empirical formulas in the molecular formula.

$$\frac{1013.7 \text{ g in molecular formula}}{506.9 \text{ g in empirical formula}} = 2.000$$

So, the molecular formula is 2 times the empirical formula. The molecular formula is $2(C_2H_3As_3Cu_2O_8)$ or $C_4H_6As_6Cu_4O_{16}$.

22. In this problem, we are asked to find the molecular formula given the molar mass and the percent composition. To determine the molecular formula, we must first find the empirical formula.

$$39.99 \text{ g C} \times \frac{1 \text{ mol C}}{12.01 \text{ g C}} = 3.330 \text{ mol C}$$

$$6.713 \text{ g H} \times \frac{1 \text{ mol H}}{1.008 \text{ g H}} = 6.660 \text{ mol H}$$

$$53.29 \text{ g O} \times \frac{1 \text{ mol O}}{16.00 \text{ g O}} = 3.331 \text{ mol O}$$

Divide each molar quantity by the smallest number of moles to convert the number of moles to a whole number.

$$\frac{3.330 \text{ mol C}}{3.330} = 1.000$$

$$\frac{6.660 \text{ mol H}}{3.330} = 2.000$$

$$\frac{3.331 \text{ mol O}}{3.330} = 1.000$$

The empirical formula is CH_2O. The molar mass of the molecule is 210.18 g. So we need to know the molar mass of the empirical formula.

$$\text{molar mass empirical formula} = 12.01 \text{ g C} + (2 \times 1.008 \text{ g H}) + 16.00 \text{ g O} = 30.03 \text{ g}$$

How many empirical formulas are there in one molecular formula? We can tell by dividing the molar mass of the molecular formula by the molar mass of the empirical formula.

$$\frac{210.18 \text{ g in molecular formula}}{30.03 \text{ g in empirical formula}} = 6.999$$

The molecular formula is 7 times the empirical formula.

$$\text{molecular formula} = 7 \times (CH_2O) \text{ or } C_7H_{14}O_7$$

CHAPTER 9

Chemical Quantities

INTRODUCTION

In this chapter you will perform many chemical calculations, all of which are based on fundamental principles such as balanced equations. A balanced equation can provide more information than is apparent at first glance. You can use a balanced equation to help answer such questions as "How much is produced?" and "How much would be needed to make this amount?" Only a balanced equation will provide correct answers to these questions.

Often, when two reactants are mixed together, one of them will run out before the other one is all used up. In a situation like this, the amount of product you can make will be limited by the reactant that is used up first. The balanced equation will help you determine which reactant runs out and how much product you can make.

CHAPTER DISCUSSION

A main point of this chapter is to be able to calculate the mass of reactant needed or mass of product formed given the mass of one or more reactants or products (this is called stoichiometry). Realize that there is really nothing new in this chapter, but you are expected to put together what you have learned in the last few chapters. For example, you need to know how to balance an equation and what it means; you need to understand the mole concept; you need to be able to calculate molar masses of molecules. If you are having difficulty with any of these, make sure to get help (and re-read these chapters) or you will find this chapter quite difficult.

Also, make sure not to get lost in the math with stoichiometry problems. By now you should be used to thinking about these problems at a molecular level. Make sure you understand what you are solving for and how you are getting there. Look back to the percent by mass problem at the end of the Chapter Discussion for Chapter Eight in this *Study Guide*. Remember that dimensional analysis is not a help if you don't understand the problem (a shortcut is not a help if you get lost along the way).

In the text, "maps" are frequently used to show how to solve these problems. Don't merely memorize these, but instead make sure to understand the thinking behind them. Note that the balanced equation is always a part of any of these maps; make sure to understand why.

One way to make sense of stoichiometry is to first consider a molecular-level sketch of a reaction. For example, consider the following problem, and answer it before you read on.

> The equation for a reaction is $2S + 3O_2 \rightarrow 2SO_3$. Consider the mixture of S and O_2 in a closed container as illustrated below:
>
> 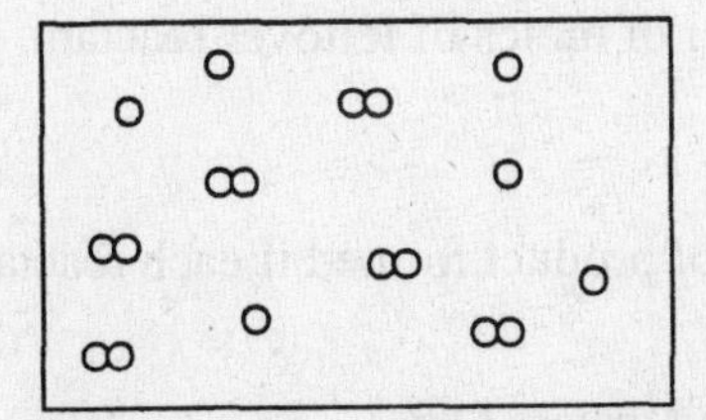
>
> ← This represents the entire container.
>
> Sketch a molecular-level representation of the product mixture.

To answer this question we need to take note of two things:

1. The number of molecules of each reactant given.
2. The ratio of the reactants that are needed.

The first piece of information is given in the problem; that is, there are six molecules of each reactant given. The second piece of information comes from the balanced equation; that is, for every two molecules of sulfur (S), three molecules of oxygen (O_2) are needed. Another way of stating this is for every two moles of sulfur (S), three moles of oxygen (O_2) are needed.

We can visualize this ratio by circling the reactants that react with each other as shown below:

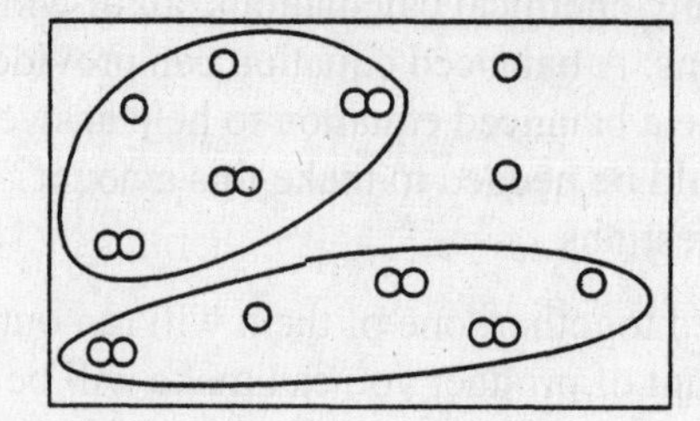

The reactants produce the compound sulfur trioxide (SO_3), and we can represent the product mixture (including any leftover reactant) as

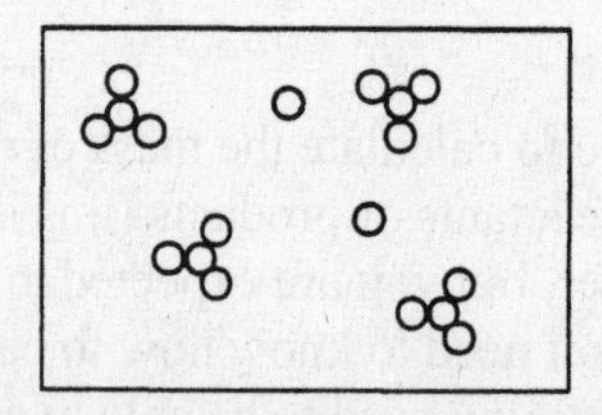

We can see that four molecules of SO_3 are produced, and two atoms of S are left over (unreacted). Make sure to understand this problem because it covers the basic concepts of stoichiometry. It even considers limiting reactants, which many students find to be the most difficult section of this chapter. The rest of stoichiometry has to do with the math (mass-mole conversions, essentially), and we will consider this briefly later.

While sketching these pictures is a good way of thinking about the problems initially, it is rather inefficient to solve all problems this way, and you will eventually want to formalize the solution a bit more. Before reading on, though, make sure to understand this example.

Formalizing a Solution

Let's consider the same problem, but different (and perhaps more efficient) ways of solving it. Recall the problem:

> You react 6 mol of S with 6 mol of O_2 according to the equation
>
> $$2S + 3O_2 \rightarrow 2SO_3$$
>
> Calculate the number of moles of SO_3 produced and the number of moles of leftover reactant.

Solution I

One way to solve this problem is to determine the number of moles of product formed if each reactant reacted completely. That is, change the given to two other problems:

1. How many moles of SO_3 could be produced from 6 mol of S and excess O_2?
2. How many moles of SO_3 could be produced from 6 mol of O_2 and excess S?

To answer this question we still need to know (as with the molecular-level sketch solution earlier) the number of moles we have (given in the problem) and the ratio from the balanced equation. Answer the two questions above before reading on.

From 6 mol of S, we can produce 6 mol of SO_3. There are a few ways to solve this with ratios, dimensional analysis, or even by inspection (the mole ratio between S and SO_3 is 2:2 or 1:1, thus for every 6 mol of S reacted, 6 mol of SO_3 will be produced).

From 6 mol of O_2, we can produce 4 mol of SO_3. Again, solve this using ratios or dimensional analysis. For example:

$$6 \text{ mol } O_2 \times \left(\frac{2 \text{ mol } SO_3}{3 \text{ mol } O_2}\right) = 4 \text{ mol } SO_3$$

So now we have two answers. That is, we have calculated 6 mol SO_3 and 4 mol SO_3. We know from our molecular level sketch that the answer is 4 mol (Note that it is NOT 10 mol—we do NOT simply add up the answers. Why not?). Let's make sense of this.

Realize what these two answers represent. They are the maximum amount of product that could be produced if the given reactant is used up completely. The answer must be the smaller number (4 mol in this case) because there is not enough O_2 to form 6 mol of SO_3. Once 4 mol of SO_3 are produced, there is no O_2 left, thus no more SO_3 can be produced. We now know, then, that 4 mol of SO_3 can be produced, and that O_2 is the limiting reactant; that is, the reactant that limits the reaction or which that runs out first.

So if all of the O_2 is used up, how much of the S is used? How much of the S is left over? We can answer these questions similarly to the previous question. We know that all 6 mol of O_2 are reacted. So how many moles of sulfur (S) would this require? Recall that the ratio between S and O_2 is 2:3 (from the balanced equation). Thus, we get

$$6 \text{ mol } O_2 \times \left(\frac{2 \text{ mol } SO_3}{3 \text{ mol } O_2}\right) = 4 \text{ mol S}$$

What does the 4 mol sulfur (S) represent? It represents the number of moles of S that are required to react with the 6 mol of O_2 that we know reacts. Thus we initially had 6 mol of sulfur, and 4 mol of S were reacted. So how many moles are left? Two moles of sulfur, just as we saw with our molecular-level sketches. This is one way of formalizing this problem–calculate the moles of product that would form if each reactant went to completion, and decide which reactant is limiting.

Solution II

There is another way we can solve this problem, and that is by comparing what we are given to what is needed for a complete reaction. For example, recall the problem

> You react 6 mol of S with 6 mol of O_2 according to the equation
>
> $$2S + 3O_2 \rightarrow 2SO_3$$
>
> Calculate the number of moles of SO_3 produced and the number of moles of leftover reactant.

We are given that we HAVE 6 mol of S and 6 mol of O_2. Can we determine which reactant is limiting without solving for the product twice?

The way to do this is to calculate the moles of each reactant that is needed to react with the moles of the other reactant that we are given. For example, we know that we have 6 mol of sulfur (S). How many moles of oxygen would be required to react completely with these 6 mol? Try this before reading on.

You should be able to calculate that 9 mol of oxygen (O_2) are needed to react with 6 mol of sulfur (S). Use the mole ratio given in the balanced equation to do so. If you are still having difficulty doing this, you need to talk with your instructor.

We know from the last solution (Solution I) that 4 mol of S are required to react with 6 mol of O_2. We can present this information in the following table:

	Moles Sulfur (S)	Moles Oxygen (O_2)
Have	6 mol S	6 mol O_2
Need	4 mol S	9 mol O_2

Note that we have more moles of sulfur than we need and fewer moles of O_2 than we need. Thus, O_2 must be the limiting reactant (the reactant that runs out first). We will therefore use the oxygen data to calculate the moles of product (SO_3) formed. That is

$$6 \text{ mol } O_2 \times \left(\frac{2 \text{ mol } SO_3}{3 \text{ mol } O_2}\right) = 4 \text{ moles } SO_3$$

Also, we can see from the table above that we have 6 mol of sulfur and need 4 mol of sulfur, thus 2 mol of sulfur are left over. This also agrees with our previous solution. It also agrees with our next solution, as we shall see.

Solution III

Another way of thinking about this problem is to set up a table that includes all the information shown in Solutions I and II. For example, consider the following

	2S	+	$3O_2$	→	$2SO_3$
Initial	6		6		0
Change	– ?		– ?		+ ?
End	?		?		?

Note that the 6 mol of each reactant (and no product initially) are represented in the "Initial" row. The "Change" row represents how much of each chemical reacts or is produced. The "End" row represents what constitutes the final reaction mixture.

Because we are assuming that the reaction goes to completion, we know that one (or possibly both) of the values for sulfur or oxygen must be zero (0) in the end row; that is, we "run out" of one (or possibly both) of the reactants. But which one(s)? We can decide this by realizing a crucial point:

The change row ratio has to be the same as the ratio of the coefficients in the balanced equation.

Make sure to understand this. The balanced equation represents the ratio of the reactants that react and the products that are formed, and the change row represents the same thing. Let's look, then, at the two possibilities:

	2S	+	$3O_2$	→	$2SO_3$
Initial	6		6		0
Change	–6		–9		+6
End	0		–3		6

	$2S$	+	$3O_2$	→	$2SO_3$
Initial	6		6		0
Change	–4		–6		+4
End	2		0		4

In the first table, we are assuming that all the sulfur is reacted, and in the second example we are assuming that all the oxygen is used. We have already looked at other solutions to this problem, so we know that the answers to the second table are correct. But we can see why sulfur is not limiting by looking at the first table. For all of the sulfur to react, we need three more moles of oxygen than we have. We cannot end up with a negative amount of oxygen, so the first table must be incorrect.

Using a table such as this is convenient because all of the possible information is conveyed; we now know which reactant is limiting, how much of the excess reactant is leftover, and how much product is formed. It also emphasizes an understanding of what a balanced equation means because we have to use the ratio for the balanced equation in the "change" row.

Stoichiometry Problems with Masses

Once you understand the concepts of stoichiometry, the rest is math. Most typical problems will give you mass data of reactants, for example, and ask for the mass of products formed. As an example, consider the following problem:

> Hydrogen gas (H_2) reacts with oxygen gas (O_2) to form water (H_2O). If you react 10.0 g of hydrogen gas with 10.0 g of oxygen gas, what mass of water can be produced? How much of which reactant is left over?

Try to solve this problem before reading on. Remember, think about the problem before merely plugging numbers into an equation.

One way to start is to determine the balanced equation for the reaction. We know that we will need to know the mole ratio of the reactants and products to solve this problem.

You should get: $2H_2 + O_2 \rightarrow 2H_2O$ as the balanced equation.

This equation tells us that for every 2 mol of hydrogen gas we need 1 mol of oxygen gas to make 2 mol of water. We need to know how many moles of each reactant we have. How do we do this? By now you should know how to convert from grams to moles. In this case, you should be able to calculate the following values:

moles H_2 : 4.96 moles

moles O_2 : 0.313 moles

If you are having difficulty getting these numbers, review molar mass in Chapter 8 of your text or talk with an instructor.

Now that we know the number of moles of each reactant, we can solve the questions that were asked in the problem. You should be able to calculate the following:

mass of water formed: 11.3 g

mass of hydrogen left over: 8.74 g

If you are having difficulty with this, review the various solutions to the previous problem in this *Study Guide*.

LEARNING REVIEW

1. Rewrite the equation below in terms of moles of reactants and products.

   6.022×10^{23} molecules $H_2(g)$ + 6.022×10^{23} molecules $I_2(g) \rightarrow 1.204 \times 10^{24}$ molecules $HI(g)$

2. How many moles of hydrogen gas could be produced from 0.8 mol sodium and an excess of water? Solve this problem by writing the equation using moles and by using the mole ratio for sodium and hydrogen.

   $$2Na(s) + 2H_2O(l) \rightarrow 2NaOH(aq) + H_2(g)$$

3. How many moles of aluminum oxide could be produced from 0.12 mol Al?

   $$4Al(s) + 3O_2(g) \rightarrow 2Al_2O_3(s)$$

4. How many moles of zinc chloride would be formed from the reaction of 1.38 mol Zn with HCl?

   $$Zn(s) + 2HCl(aq) \rightarrow ZnCl_2(aq) + H_2(g)$$

5. Solid silver carbonate decomposes to produce silver metal, oxygen gas and carbon dioxide.

   a. Write a balanced chemical equation for this reaction.

   b. What mass of silver will be produced by the decomposition of 6.32 g silver carbonate?

6. When aqueous solutions of sodium sulfate and lead(II) nitrate are mixed, a solid white precipitate is formed. How much solid lead(II) sulfate could be produced from 12.0 g Na_2SO_4 if $Pb(NO_3)_2$ is in excess?

   $$Na_2SO_4(aq) + Pb(NO_3)_2(aq) \rightarrow PbSO_4(s) + 2NaNO_3(aq)$$

7. Hydrogen gas and chlorine gas will combine to produce gaseous hydrogen chloride. How many molecules of hydrogen chloride can be produced from 20.1 g hydrogen gas and excess chlorine gas?

8. Some lightweight backpacking stoves use kerosene as a fuel. Kerosene is composed of carbon and hydrogen, and although it is a mixture of molecules, we can represent the formula of kerosene as $C_{11}H_{24}$. When a kerosene stove is lit, the fuel reacts with oxygen in the air to produce carbon dioxide gas and water vapor. If it takes 15 g of kerosene to fry a trout for dinner, how many grams of water are produced?

   $$C_{11}H_{24}(l) + 17O_2(g) \rightarrow 11CO_2(g) + 12H_2O(g)$$

9. You are trying to prepare 6 copies of a three-page report. If you have on hand 6 copies of pages one and two, and 4 copies of page three

   a. How many complete reports can you produce?

   b. Which page limits the number of complete reports you can produce?

10. Manganese(IV) oxide reacts with hydrochloric acid to produce chlorine gas, manganese(II) chloride and water.

    $$MnO_2(s) + 4HCl(aq) \rightarrow Cl_2(g) + MnCl_2(aq) + 2H_2O(l)$$

    a. When 10.2 g MnO_2 react with 18.3 g HCl, which is the limiting reactant?

    b. What mass of chlorine gas can be produced?

    c. How many molecules of water can be produced?

11. The acid–base reaction between phosphoric acid and magnesium hydroxide produces solid magnesium phosphate and liquid water. If 121.0 g of phosphoric acid reacts with 89.70 g magnesium hydroxide, how many grams of magnesium phosphate will be produced?

12. If 85.6 g of potassium iodide reacts with 2.41×10^{24} molecules of chlorine gas, how many grams of iodine can be produced?

$$Cl_2(g) + 2KI(s) \rightarrow 2KCl(s) + I_2(s)$$

13. Aqueous sodium iodide reacts with aqueous lead(II) nitrate to produce the yellow precipitate lead(II) iodide and aqueous sodium nitrate.

    a. What is the theoretical yield of lead iodide if 125.5 g of sodium iodide reacts with 205.6 g of lead nitrate?

    b. If the actual yield from this reaction is 197.5 g lead iodide, what is the percent yield?

ANSWERS TO LEARNING REVIEW

1. 6.022×10^{23} molecules is equivalent to 1 mol of molecules, and 1.204×10^{24} molecules is equivalent to $2(6.022 \times 10^{23}$ molecules), so the equation can be rewritten as

$$1 \text{ mol } H_2(g) + 1 \text{ mol } I_2(g) \rightarrow 2 \text{ mol } HI(g)$$

2. The balanced equation tells us that two moles of sodium react with two moles of water to form two moles of sodium hydroxide and four moles of hydrogen. By using mole ratios determined from the balanced equation, we can calculate the number of moles of reactants required and products produced from 0.8 mol sodium.

$$0.8 \text{ mol Na} \times \frac{2 \text{ mol } H_2O}{2 \text{ mol Na}} = 0.8 \text{ mol } H_2O$$ 0.8 mol sodium requires 0.8 mol H_2O.

$$0.8 \text{ mol Na} \times \frac{2 \text{ mol NaOH}}{2 \text{ mol Na}} = 0.8 \text{ mol NaOH}$$ 0.8 mol Na produces 0.8 mol NaOH.

$$0.8 \text{ mol Na} \times \frac{1 \text{ mol } H_2}{2 \text{ mol Na}} = 0.4 \text{ mol } H_2$$ 0.8 mol Na produces 0.4 mol H_2.

We can write the molar values we have calculated in equation form.

$$0.8 \text{ mol Na}(s) + 0.8 \text{ mol } H_2O(l) \rightarrow 0.8 \text{ mol NaOH}(aq) + 0.4 \text{ mol } H_2(g)$$

3. First make sure the equation is balanced. You should always determine whether or not an equation is balanced, and balance it if necessary. To solve this problem, we need to know the mole ratio for aluminum and aluminum oxide. The mole ratio represents the relationship between the mol of substance given in the problem and the mol of the desired substance and is taken directly from the balanced equation. The mole ratio for aluminum oxide and aluminum is.

$$\frac{2 \text{ mol } Al_2O_3}{4 \text{ mol Al}}$$

$$0.12 \text{ mol Al} \times \frac{2 \text{ mol } Al_2O_3}{4 \text{ mol Al}} = 0.060 \text{ mol } Al_2O_3$$

4. First make sure the equation is balanced. The mole ratio for zinc and zinc chloride is taken from the balanced equation and is.

$$\frac{1 \text{ mol ZnCl}_2}{1 \text{ mol Zn}}$$

$$1.38 \text{ mol Zn} \times \frac{1 \text{ mol ZnCl}_2}{1 \text{ mol Zn}} = 1.38 \text{ mol ZnCl}_2$$

Because there is a 1:1 mole ratio of $ZnCl_2$ to Zn, the number of moles of zinc equals the moles of zinc chloride produced.

5.

a. First write the formulas for reactants and products. Include the physical states. Then balance the equation.

$$2Ag_2CO_3(s) \rightarrow 4Ag(s) + O_2(g) + 2CO_2(g)$$

b. It is not possible to solve this problem by converting directly from grams of Ag_2CO_3 to grams of Ag. However, the balanced equation tells us the relationship between Ag_2CO_3 and Ag in moles. If we can convert grams of Ag_2CO_3 to moles, we can use the mole ratio to tell us how many moles of Ag are produced. To convert grams of Ag_2CO_3 to moles, you can produce a conversion factor from the equivalence statement that relates the number of moles to molar mass. The correct conversion factor is.

$$\frac{1 \text{ mol Ag}_2\text{CO}_3}{275.75 \text{ g Ag}_2\text{CO}_3}$$

$$6.32 \text{ g Ag}_2\text{CO}_3 \times \frac{1 \text{ mol Ag}_2\text{CO}_3}{275.75 \text{ g Ag}_2\text{CO}_3} = 0.0229 \text{ mol Ag}_2\text{CO}_3$$

Now we can use the mole ratio for Ag_2CO_3 and Ag to calculate the moles of Ag.

$$0.0229 \text{ mol Ag}_2\text{CO}_3 \times \frac{4 \text{ mol Ag}}{2 \text{ mol Ag}_2\text{CO}_3} = 0.0458 \text{ mol Ag}$$

We now know the moles of Ag, but we want to know the grams of Ag. The conversion factor below, which is derived from the molar mass of silver, will allow us to calculate grams.

$$\frac{107.87 \text{ g Ag}}{1 \text{ mol Ag}}$$

$$0.0458 \text{ mol Ag} \times \frac{107.87 \text{ g Ag}}{1 \text{ mol Ag}} = 4.94 \text{ g Ag}$$

If we string together all the parts of this problem, we can see that the overall strategy is to convert grams to moles using the molar mass, then moles to moles using the mole ratio, and moles to mass using the molar mass.

$$6.32 \text{ g Ag}_2\text{CO}_3 \times \frac{1 \text{ mol Ag}_2\text{CO}_3}{175.75 \text{ g Ag}_2\text{CO}_3} \times \frac{4 \text{ mol Ag}}{2 \text{ mol Ag}_2\text{CO}_3} \times \frac{107.87 \text{ g Ag}}{1 \text{ mol Ag}} = 4.94 \text{ g Ag}$$

↑	↑	↑	↑	↑
grams of reactant	molar mass of reactant	mole ratio	molar mass of product	grams of product

6. This question provides us with grams of reactant and asks for grams of product. Because we are told that $Pb(NO_3)_2$ is in excess, the limiting reactant must be Na_2SO_4. The amount of precipitate that can be formed is determined by the amount of Na_2SO_4 relative to the grams of $PbSO_4$. We must first calculate the moles of Na_2SO_4, then use the mole ratio derived from the balanced equation to tell us how many moles of $PbSO_4$ are produced, and finally, we can use the molar mass of $PbSO_4$ to calculate the grams of $PbSO_4$.

$$12.0 \text{ g } Na_2SO_4 \times \frac{1 \text{ mol } Na_2SO_4}{142.05 \text{ g } Na_2SO_4} \times \frac{1 \text{ mol } PbSO_4}{1 \text{ mol } Na_2SO_4} \times \frac{303.27 \text{ g } PbSO_4}{1 \text{ mol } PbSO_4} = 25.6 \text{ g } PbSO_4$$

7. First write the balanced equation for this reaction.

$$H_2(g) + Cl_2(g) \rightarrow 2HCl(g)$$

This problem gives us grams of hydrogen and asks for molecules of hydrogen chloride. There is no way to convert grams of hydrogen directly to molecules of hydrogen chloride. However, we can convert grams of hydrogen to moles of hydrogen using the molar mass of hydrogen gas. The balanced equation provides a mole ratio so that we can calculate the moles of hydrogen chloride. Converting from moles to molecules can be done because we know that 1 mol of hydrogen chloride equals 6.022×10^{23} molecules of hydrogen chloride.

$$20.1 \text{ g } H_2 \times \frac{1 \text{ mol } H_2}{2.016 \text{ g } H_2} \times \frac{2 \text{ mol HCl}}{1 \text{ mol } H_2} \times \frac{6.022 \times 10^{23} \text{ molecules HCl}}{1 \text{ mol HCl}}$$
$$= 1.20 \times 10^{25} \text{ molecules}$$

8. We are given grams of kerosene and asked for grams of water vapor. Because we cannot convert directly between grams of kerosene and grams of water, we first convert grams of kerosene to moles of kerosene using the molar mass of kerosene. Then use the mole ratio of kerosene and water from the balanced equation to determine the moles of water vapor. The molar mass of water will allow us to convert moles to grams of water.

$$15 \text{ g } C_{11}H_{24} \times \frac{1 \text{ mol } C_{11}H_{24}}{156.30 \text{ g } C_{11}H_{24}} \times \frac{12 \text{ mol } H_2O}{1 \text{ mol } C_{11}H_{24}} \times \frac{18.02 \text{ g } H_2O}{1 \text{ mol } H_2O} = 21 \text{ g } H_2O$$

9.

a. You can prepare 4 complete copies. Copies 5 and 6 would lack page three.

b. Page three limits the number of complete reports that can be produced.

10.

a. By looking at the grams of MnO_2 and the grams of HCl, it is impossible to tell which is the limiting reactant. It is possible to compare moles of reactants because we know the mole ratio of reactants from the balanced equation. So calculate the number of moles of each reactant. Then determine how many moles of product could be produced from each of the two reactants. The reactant that allows the fewest number of moles of product is the limiting reactant.

$$10.2 \text{ g } MnO_2 \times \frac{1 \text{ mol } MnO_2}{86.94 \text{ g } MnO_2} \times \frac{1 \text{ mol } Cl_2}{1 \text{ mol } MnO_2} = 0.117 \text{ mol } Cl_2$$

$$18.3 \text{ g HCl} \times \frac{1 \text{ mol HCl}}{36.46 \text{ g HCl}} \times \frac{1 \text{ mol } Cl_2}{4 \text{ mol HCl}} = 0.125 \text{ mol } Cl_2$$

From 10.2 MnO_2, 0.117 mol Cl_2 can be produced, and from 18.3 g HCl, 0.125 mol Cl_2 can be produced. So the limiting reactant is MnO_2.

b. We already know that the most chlorine we can make is 0.117 mol. To convert from moles to grams, use the molar mass of a chlorine molecule.

$$0.117 \text{ mol } Cl_2 \times \frac{70.90 \text{ g } Cl_2}{1 \text{ mol } Cl_2} = 8.30 \text{ g } Cl_2$$

c. The limiting reactant is manganese(IV) oxide, so we need to calculate the moles of water that can be produced from 10.2 g MnO_2. By using the mole ratio from the balanced equation we can calculate the moles of water. To convert from moles of water to the number of molecules, use Avogadro's number as a conversion factor.

$$10.2 \text{ g } MnO_2 \times \frac{1 \text{ mol } MnO_2}{86.94 \text{ g } MnO_2} \times \frac{2 \text{ mol } H_2O}{1 \text{ mol } MnO_2} \times \frac{6.022 \times 10^{23} \text{ molecules } H_2O}{1 \text{ mol } H_2O}$$
$$= 1.41 \times 10^{23} \text{ molecules } H_2O$$

11. First write the balanced equation for this reaction.

$$2H_3PO_4(aq) + 3Mg(OH)_2(s) \rightarrow Mg_3(PO_4)_2(s) + 6H_2O(l)$$

When the mass is given for two reactants,, and you are asked to determine the quantity of product that can be produced, you must first determine which reactant is limiting. Determine how many moles of product would be produced from each reactant. The reactant that will produce the fewest number of moles of product is the limiting reactant.

$$121.0 \text{ g } H_3PO_4 \times \frac{1 \text{ mol } H_3PO_4}{97.99 \text{ g } H_3PO_4} \times \frac{1 \text{ mol } Mg_3(PO_4)_2}{2 \text{ mol } H_3PO_4} = 0.6174 \text{ mol } Mg_3(PO_4)_2$$

$$89.70 \text{ g } Mg(OH)_2 \times \frac{1 \text{ mol } Mg(OH)_2}{58.33 \text{ g } Mg(OH)_2} \times \frac{1 \text{ mol } Mg_3(PO_4)_2}{3 \text{ mol } Mg(OH)_2} = 0.5126 \text{ mol } Mg_3(PO_4)_2$$

In this reaction, the $Mg(OH)_2$ is the limiting reactant. We now know how many moles of $Mg_3(PO_4)_2$ are produced, but we want to know the number of grams. Use the molar mass of $Mg(PO_4)_2$ to convert from moles of grams.

$$0.5126 \text{ mol } Mg_3(PO4)_2 \times \frac{262.87 \text{ g } Mg_3(PO4)_2}{1 \text{ mol } Mg_3(PO4)_2} = 134.7 \text{ g } Mg_3(PO4)_2$$

12. In this problem we are given quantities of two reactants, one expressed in grams and the other in molecules. Before we can calculate grams of product, we need to know which reactant limits the amount of product that can be produced. Convert the grams of KI to moles using the molar mass of KI, and calculate the moles of I_2 from the mole ratio.

$$85.6 \text{ g KI} \times \frac{1 \text{ mol KI}}{166.0 \text{ g KI}} \times \frac{1 \text{ mol } I_2}{2 \text{ mol KI}} = 0.258 \text{ mol } I_2$$

The quantity of the other reactant, Cl_2, is given in molecules, not grams. We can convert molecules of Cl_2 to moles of Cl_2 using Avogadro's number, 1 mol Cl_2 = 6.022×10^{23} molecules Cl_2.

$$2.41 \times 10^{24} \text{ molecules } Cl_2 \times \frac{1 \text{ mol } Cl_2}{6.022 \times 10^{23} \text{ molecules } Cl_2} = 4.00 \text{ mol } Cl_2$$

So 2.41×10^{24} molecules is equivalent to 4.00 moles of Cl_2. Now we can find the moles of I_2 that can be produced from 4.00 moles of Cl_2.

$$4.00 \text{ mol } Cl_2 \times \frac{1 \text{ mol } I_2}{1 \text{ mol } Cl_2} = 4.00 \text{ mol } I_2$$

KI limits the amount of I_2 that can be produced, so it is the limiting reactant. We can calculate the grams of I_2 using the molar mass of I_2.

$$0.258 \text{ mol } I_2 \times \frac{253.8 \text{ g } I_2}{1 \text{ mol } I_2} = 65.5 \text{ g } Cl_2$$

13.

a. First balance the equation.

$$2 \text{ NaI} + Pb(NO_3)_2 \rightarrow PbI_2 + 2 \text{ NaNO}_3$$

This problem first asks for the theoretical yield of PbI_2 when two quantities of reactants are mixed. Before we can calculate the amount of product we need to know which reactant is limiting. Use the molar mass for each product and the mole ratio for the balanced equation to calculate the moles of PbI_2 that could be produced.

$$125.5 \text{ g NaI} \times \frac{1 \text{ mol NaI}}{149.89 \text{ g NaI}} \times \frac{1 \text{ mol } PbI_2}{2 \text{ mol NaI}} = 0.4186 \text{ mol } PbI_2$$

$$205.6 \text{ g } Pb(NO_3)_2 \times \frac{1 \text{ mol } Pb(NO_3)_2}{331.22 \text{ g } Pb(NO_3)_2} \times \frac{1 \text{ mol } PbI_2}{1 \text{ mol } Pb(NO_3)_2} = 0.6207 \text{ mol } PbI_2$$

The limiting reactant is NaI. Now we can answer the question about theoretical yield. Theoretical is the amount of product we calculate can be produced, that is, 0.4186 mol PbI_2. In real life, the actual yield might be less than the calculated yield. The theoretical yield of PbI_2 can be calculated from the number of moles of PbI_2, if we know the molar mass.

$$0.4186 \text{ mol } PbI_2 \times \frac{461.00 \text{ g } PbI_2}{1 \text{ mol } PbI_2} = 193.0 \text{ g } PbI_2$$

b. In part a we calculated the theoretical yield of lead(II) iodide, which is 193.0 g. We are told that the actual yield from this reaction was found to be 164.5 g. The percent yield is equal to the actual yield divided by the theoretical yield, multiplied by 100 percent. So the percent yield of lead(II) iodide is

$$\frac{164.5 \text{ g } PbI_2}{193.0 \text{ g } PbI_2} \times 100\% = 85.23\%$$

CHAPTER 10

Energy

INTRODUCTION

Chemistry is about matter and its changes. As a part of all chemical reactions, energy (in the form of heat) is either released or required. Even though energy is converted from one form to another as a chemical reaction proceeds, the total amount of energy in the universe is constant (this is known as the law of conservation of energy). In this chapter you will learn about the nature of energy and energy sources, discover how we can predict whether or not a reaction will be spontaneous, and calculate the amount of energy needed to heat water and other substances.

CHAPTER DISCUSSION

Energy is defined as the ability to do work or produce heat. For example, imagine that we move a chair across a room. We do work on the chair because we need to exert a force to get the chair to move, and we move it over a certain distance (work is defined as a force acting over a distance). The heavier the chair, the more force required, so the more work is done (or the more energy is required). The farther we move the chair (the larger the distance), the more work is done, and the more energy is required. Heat is flow of energy due to a temperature difference. Although we often talk of heat as though it were a substance ("Close the window, you are letting the heat out"), heat is not a "thing."

Another useful way of thinking about energy is the following:

Energy is what is required in order to resist a natural tendency.

For example, consider holding a bowling ball above your head. You are not moving the ball, so there is no work. But obviously you get tired after awhile, so it feels as though you are exerting energy. How can this be? The natural tendency of the bowling ball is to fall to the ground (due to gravity). By keeping the ball from falling, you are making the ball resist its natural tendency; therefore you *are* exerting energy. The same goes if you are holding a string connected to a helium balloon. The natural tendency of the balloon is to float away, and by holding the string you are keeping the balloon from doing so. This, too, requires energy.

In discussing energy, you also need to keep distinct potential energy and kinetic energy. The "ball on a hill" example given in the text is a good one to read and understand, but it is a non-chemical example. When thinking about chemistry, keep in mind that the potential energy of a chemical system is stored in the bonds (you will learn more about chemical bonds in Chapter 12, but it is enough for now to know that a chemical bond is a force that holds atoms together in a molecule). In order to "break apart" a molecule, energy is required to break the bonds. As bonds reform when new molecules are made, energy is released in the form of heat (as kinetic energy).

Let's look at the reaction that occurs in a Bunsen burner (the combustion of methane):

$$CH_4(g) + 2O_2(g) \rightarrow CO_2(g) + H_2O(l).$$

In this case, more energy is released when the bonds form CO_2 and H_2O than is required to break the bonds in CH_4 and O_2. Thus this reaction is exothermic, and we report the heat with a negative sign (in this case ($\Delta H = -891$ kJ/mol). See Figure 10.5 in your text for a potential energy diagram of this reaction. It would be a good idea to see if you can draw a similar diagram for an endothermic reaction.

You should also be able to differentiate between heat and temperature. Specifically, make sure you understand that heat and temperature are not the same. Temperature is a measure of the random motions of the particles that make up a substance. The concept of specific heat capacity helps us see this difference. You undoubtedly have seen that different substances change temperatures differently when the same amount of heat is transferred. For example, if you are making soup on the stove and are stirring with a metal spoon, you notice that the spoon gets hot rather quickly. A wooden spoon does not get nearly so hot, even though the amount of heat is the same. Why is this? Different substances, due to their make-ups, react to heat differently. We quantify this with the specific heat capacity, which is defined as the amount of energy required to change the temperature of one gram of the substance by one Celsius degree. See Table 10.1 in your text for a list of specific heat capacities, and notice that the heat capacities of metals are lower than water. This means that the temperature of a given mass of metal will increase much more than the temperature of the same mass of water when the same amount of heat is transferred to each. Practice doing problems dealing with specific heat capacities to make sure you can either use heat capacities to calculate temperature differences or that you can calculate heat capacities and determine the substance that is being heated or cooled.

You will also be expected to calculate the heat (enthalpy) of a reaction from known heats of related reactions. You can do this using Hess's law, which works because energy is a state function. Make sure you understand how Hess's law uses the idea that energy is a state function (another way of stating this is to rely on the existence of the first law of thermodynamics – make sure you know why).

While energy is conserved (that is the quantity stays the same), the quality of energy is constantly decreasing. That is, the amount of usefulness in a sample of energy decreases as the energy is "used." The example used in your text is gasoline. A sample of gasoline can be thought of as concentrated energy (potential energy). However, as you drive the car heat is released to the road, the air, etc., and so, while the amount of energy in the universe is the same before and after you drive your car, the usefulness has decreased. Natural processes always occur in a way that increases the "spreading" of energy and thus decreases the usefulness of the energy. We term this entropy, which is a measure of disorder. For a process to be spontaneous, the entropy of the universe must increase.

LEARNING REVIEW

1. Explain the difference between *kinetic energy* and *potential energy*.
2. Why isn't all energy available as work?
3. The law of conservation of energy means that energy is a state function. Explain why.
4. Explain differences among heat, temperature, and thermal energy.
5. Provide a molecular-level explanation of why the temperatures of a cold soft drink and hot coffee in the same room will eventually be the same.
6. In which case is more heat involved: mixing 100.0-g samples of 90 °C water and 80 °C water or mixing 100.0-g samples of 60 °C water and 10 °C water? Assume no heat is lost to the environment.
7. What is meant by potential energy in a chemical reaction? Where is it located?
8. Are the following processes exothermic or endothermic?
   a. When solid KBr is dissolved in water, the solution gets colder.
   b. Natural gas (CH_4) is burned in a furnace.
   c. When concentrated sulfuric acid is added to water, the solution gets very hot.
   d. Water is boiled in a tea kettle.

9. In thermodynamics the chemist takes the system's point of view. What does this statement mean?

10. A gas absorbs 45 kJ of heat and does 29 kJ of work. Calculate ΔE.

11. Convert the energy values below to the desired units.
    a. 45.8 cal to J
    b. 0.561 cal to J
    c. 5.96 J to cal
    d. 76 J to cal

12. Calculate the number of calories required to change the temperature of each of the quantities of water below.
    a. 100.1 g of water from 6°C to 25°C
    b. 2.32 g of water from 36°F to 42°F
    c. 40 g of water by 12°C
    d. 16.9 g of water from 75.0°C to 80.0°C

13. How much energy (in joules) is required to raise the temperature of 25.2 g of solid carbon rod from 25 °C to 50.°C? The specific heat capacity of solid carbon is 0.71 J/g °C.

14. How much energy (in calories) is required to raise the temperature of 10. g steam from 122.2 °C to 130.4 °C? The specific heat capacity of water(*g*) is 2.0 J/g °C.

15. How much of a temperature change would occur if 2736.8 J of energy were applied to a piece of iron bar weighing 450.5 g? The specific heat capacity of solid iron is 0.45 J/g °C.

16. What is the mass in grams of a piece of aluminum wire if a change in temperature of 5.67 °C required 8.53 J? The specific heat capacity of solid aluminum is 0.89 J/g °C.

17. What is the specific heat capacity of ethyl alcohol if 1972.4 J of energy is necessary to raise the temperature of 53.4 g ethyl alcohol by 15.2°C?

18. Calculate the enthalpy change when 1.00 g of methane is burned in excess oxygen according to the reaction $CH_4(g) + 2O_2(g) \rightarrow CO_2(g) + H_2O(l)$ ($\Delta H = -891$ kJ/mol).

19. Given the following data:

$$C_2H_2(g) + \tfrac{5}{2}O_2(g) \rightarrow 2CO_2(g) + H_2O(l) \qquad \Delta H = -1300.\ \text{kJ/mol}$$
$$C(s) + O_2(g) \rightarrow CO_2(g) \qquad \Delta H = -394\ \text{kJ/mol}$$
$$H_2(g) + \tfrac{1}{2}O_2(g) \rightarrow H_2O(l) \qquad \Delta H = -286\ \text{kJ/mol}$$

calculate ΔH for the reaction

$$2C(s) + H_2(g) \rightarrow C_2H_2(g)$$

20. What is the difference between the *quality* of energy and the *quantity* of energy? Which is decreasing?

21. Which energy sources used in the United States have declined the most in the last 150 years? Which have increased the most?

22. Why can't the first law of thermodynamics explain why a ball doesn't spontaneously roll up a hill?

23. Exothermic reactions have a driving force. Nevertheless, water melting into a liquid is endothermic, and this process occurs at room temperature. Explain why.

ANSWERS TO LEARNING REVIEW

1. Kinetic energy is the energy of motion. Potential energy is the energy of position.
2. Some energy is given off in other forms such as heat or light.
3. Because energy is conserved in any process, the pathway of the process does not matter. This means that energy is a state function.
4. Heat is a flow of energy due to a temperature difference; temperature is a measure of the average kinetic energy of a substance; thermal energy comes from the random motion of the components of the system.
5. Temperature is a measure of the average kinetic energy of the samples. The "coffee particles" (mostly water molecules) are of higher average kinetic energy than the "air particles" (a mixture of mostly nitrogen and oxygen molecules) in the room, which are of higher energy than the "soft drink particles" (mostly water molecules). At the coffee-air interface, a collision of higher-energy "coffee particle" and "air particle" results in energy being transferred from the coffee to the room. Transfer also occurs from air to soft drink because "air particles" are of higher energy than the "soft drink particles" (due to the temperature difference). The energy transfers result in the eventual average kinetic energies of each sample being equal, which means the temperatures are equal. Because the volume of air is so large (the system is open), however, no noticeable temperature change of the air will result.
6. There is more heat involved in mixing 100.0-g samples of 60 °C and 10 °C water because there is a large temperature difference (and heat is a flow of energy due to a temperature difference).
7. The potential energy is the energy available to do work. Potential energy in a chemical reaction is stored in the chemical bonds.
8.
   a. endothermic
   a. exothermic
   b. exothermic
   c. endothermic
9. The chemist chooses the sign based on whether energy flows from the system (negative sign) or into the system (positive sign).
10. $\Delta E = q + w = 45\text{ kJ} + (-29\text{ kJ}) = 16\text{ kJ}$
11. Converting calories to joules or joules to calories requires knowing that 1 cal = 4.184 J.
    a. $45.8\text{ cal} \times \frac{4.184\text{ J}}{1\text{ cal}} = 192\text{ J}$
    b. $0.561\text{ cal} \times \frac{4.184\text{ J}}{1\text{ cal}} = 2.35\text{ J}$
    c. $5.96\text{ J} \times \frac{1\text{ cal}}{4.184\text{ J}} = 1.42\text{ cal}$

d. $76\text{ J} \times \dfrac{1\text{ cal}}{4.184\text{ J}} = 18\text{ cal}$

12. These problems ask you to calculate the calories required to heat a quantity of water. One calorie is defined as the amount of heat required to raise the temperature of 1 gram of water by 1 degree Celsius. To solve these problems, you need to multiply the number of grams of water to be heated by the number of degrees Celsius change in the temperature of the water.

a. $100.1\text{ g water} \times 19°\text{C} \times \dfrac{1\text{ cal}}{\text{g water} \times °\text{C}} = 1900\text{ cal}$

b. The initial and final temperatures of water are given in °F. We must convert to °C before solving the problem.

Initial temperature:

$$T_{°C} = \frac{(T_{°F} - 32)}{1.80}$$

$$T_{°C} = \frac{(36 - 32)}{1.80}$$

$$T_{°C} = 2.2$$

Final temperature:

$$T_{°C} = \frac{(T_{°F} - 32)}{1.80}$$

$$T_{°C} = \frac{(42 - 32)}{1.80}$$

$$T_{°C} = 5.6$$

Temperature change:

$$5.6\ °\text{C} - 2.2\ °\text{C} = 3.4\ °\text{C}$$

Solution: $2.32\text{ g water} \times 3.3°\text{C} \times \dfrac{1\text{ cal}}{\text{g water} \times °\text{C}} = 7.9\text{ cal}$

c. $40.\text{ g water} \times 12°\text{C} \times \dfrac{1\text{ cal}}{\text{g water} \times °\text{C}} = 480\text{ cal}$

d. $16.9\text{ g water} \times 5.0°\text{C} \times \dfrac{1\text{ cal}}{\text{g water} \times °\text{C}} = 85\text{ cal}$

13. In this problem we want to calculate the heat energy needed to raise the temperature of a substance other than water. To do this, we need to know the specific heat capacity of the substance. The specific heat capacity tells us the amount of heat energy required to change the temperature of 1 gram of a substance by 1 degree Celsius. Every substance has its own specific heat capacity. That of solid carbon is 0.71 J/g °C.

If it takes 0.71 J to raise the temperature of 1 gram of carbon 1 degree Celsius, then it will take 0.71 J × 25.2 to raise 25.2 g carbon by 1 degree Celsius. We wish to raise the temperature of the carbon rod by 25 °C, not 1 °C. We will need twenty-five times the heat energy needed to raise the temperature of 25.2 g carbon by 1 degree Celsius.

$$\text{Joules} = \frac{0.71\ \text{J}}{\text{g carbon} \times {}^\circ\text{C}} \times 25.2\ \text{g carbon} \times 25^\circ\text{C}$$

The joules required to raise the temperature of 25.2 g carbon by 25 °C = 450 J.

14. When you are given the number of grams of a substance, a change in temperature in degrees Celsius, and the specific heat capacity for that substance and are asked to calculate the heat energy required, you can use the formula $Q = s \times m \times \Delta T$. The specific heat capacity is s, m equals the mass in grams, ΔT is the change in temperature in degrees Celsius, and Q is the heat energy required. We can solve this problem with the formula although we will need to convert Q from joules to calories since calories are asked for.

$$Q = s \times m \times \Delta T$$

$$Q = \frac{2.0\ \text{J}}{\text{g water}(g) \times {}^\circ\text{C}} \times 10.\ \text{g water}(g) \times 8.2\ {}^\circ\text{C}$$

$$Q = 160\ \text{J}$$

The answer should be expressed in calories:

$$160\ \text{J} \times \frac{1\ \text{cal}}{4.184\ \text{J}} = 38\ \text{cal}$$

15. In this problem, we are given Q, the heat energy in joules; m, the mass in grams of a piece of iron; and s, the specific heat capacity of iron. We are asked for ΔT, the change in temperature. If we rearrange the equation $Q = s \times m \times \Delta T$, we can solve for ΔT.

Divide both sides by ΔT.

$$\frac{Q}{\Delta T} = s \times m$$

Now, divide both sides by Q (same as multiplying by $\frac{1}{Q}$).

$$\frac{Q}{\Delta T} \times \frac{1}{Q} = \frac{s \times m}{Q}$$

We now have $1/\Delta T$ (the inverse of ΔT) isolated on one side of the equation.

$$\frac{1}{\Delta T} = \frac{s \times m}{Q}$$

Invert both sides of the equation.

$$\frac{\Delta T}{1} = \frac{Q}{s \times m}$$

$$\frac{\Delta T}{1} = \Delta T = \frac{Q}{s \times m}$$

Now, find ΔT.

$$\Delta T = \frac{2736.8 \text{ J}}{\frac{0.45 \text{ J}}{\text{g °C}} \times 450.5 \text{ g}}$$

$$\Delta T = 14 \text{ °C}$$

16. **In this problem we are asked to solve for mass, m. We are given ΔT, Q, and s. We can rearrange the equation as illustrated below.**

    Divide both sides of the equation by m.

$$\frac{Q}{m} = \frac{s \times m \times \Delta T}{m}$$

Divide both sides of the equation by Q.

$$\frac{Q}{m} \times \frac{1}{Q} = s \times \Delta T \times \frac{1}{Q}$$

$$\frac{1}{m} = \frac{s \times \Delta T}{Q}$$

Invert both sides of the equation.

$$\frac{m}{1} = m = \frac{Q}{s \times \Delta T}$$

Now, find m.

$$m = \frac{8.53 \text{ J}}{\frac{0.89 \text{ J}}{\text{g} \times \text{°C}} \times 5.67 \text{ °C}}$$

$$m = 1.7 \text{ g}$$

17. **We are asked to find the specific heat capacity, s, when given ΔT, Q, and m. Rearrange the equation to isolate s on one side of the equation.**

$$s = \frac{Q}{m \times \Delta T}$$

Now substitute values into the equation.

$$s = \frac{1972.4 \text{ J}}{54 \text{ g} \times 15.2 \text{ °C}} = 2.43 \text{ J/g °C}.$$

The specific heat capacity of ethyl alcohol is 2.43 J/g °C.

18. $1.00 \text{ g } CH_4 \times \frac{1 \text{ mol } CH_4}{16.04 \text{ g } CH_4} \times \frac{-891 \text{ kJ}}{1 \text{ mol } CH_4} = -55.5 \text{ kJ}$

$2C(s) + 2O_2(g) \rightarrow 2CO_2(g)$

$\Delta H = 2(-394 \text{ kJ/mol})$

$H_2(g) + \frac{1}{2}O_2(g) \rightarrow H_2O(l)$

$\Delta H = -286 \text{ kJ/mol}$

$2CO_2(g) + H_2O(l) \rightarrow C_2H_2(g) + \frac{5}{2}O_2(g)$

$\Delta H = -(-1300. \text{ kJ/mol})$

$2C(s) + H_2(g) \rightarrow C_2H_2(g) \qquad \Delta H = 226 \text{ kJ/mol}$

19. The quality of energy tells us the form of the energy (potential or kinetic). The quantity of energy tells us how much. The total amount (or quantity) is conserved. However, when potential energy is converted to kinetic energy, we say the quality is decreasing.

20. Wood has decreased the most. Nuclear and petroleum/natural gas have increased the most (see Figure 10.7).

21. The first law of thermodynamics tells us the total amount of energy is constant. It does not tell us anything about the direction of energy transfer.

22. The energy of the surroundings is used to melt solid water. The temperature of the water will eventually reach that of the room, and room temperature is higher than the melting point of water. In addition, liquid water has greater entropy than solid water, so entropy is the driving force in this process.

CHAPTER 11

Modern Atomic Theory

INTRODUCTION

It is difficult to form a mental image of atoms because we can't see them. Scientists have produced models that account for the behavior of atoms by making observations about the properties of atoms. So we know quite a bit about these tiny particles which make up matter even though we cannot see them. In this chapter you will learn how chemists believe atoms are structured.

CHAPTER DISCUSSION

This is yet another chapter in which models play a significant role. Recall that we like models to be simple, but we also need models to explain the questions we want to answer. For example, Dalton's model of the atom has the advantage of being quite simple and is useful when considering molecular-level views of solids, liquids, and gases and in representing chemical equations. However, it cannot explain fundamental questions such as why atoms "stick" together to form molecules. Scientists such as Thomson and Rutherford expanded the model of the atom to include subatomic particles. The model is more complicated than Dalton's, but it begins to explain chemical reactivity (electrons are involved) and formulas for ionic compounds. But questions still abound. For example, why are there similarities in reactivities of elements (periodic trends)?

To begin our understanding of modern atomic theory, let's first discuss some observations. For example, you have undoubtedly seen a fireworks display, either in person or on television. Where do the colors come from? How do we get so many different colors? It turns out that different salts, when heated, give off different colors. For example, copper(II) chloride is a characteristic green color, and lithium chloride is red. But why?

To answer this, let's consider a light bulb and why we see white light from it. The light that we see is part of a spectrum of electromagnetic radiation, which includes x-rays, UV rays, visible light, microwaves, etc. (see Section 11.2 in your text). A light bulb lights up because a thin filament in the bulb is heated (electric current is sent through the thin filament), and this heat (energy) is released as light. Realize that energy is related to the wavelength of electromagnetic radiation, and the wavelength (if visible) is related to color. Because the wire in the light bulb is heated, energy of all wavelengths is emitted, and therefore wavelengths of all colors are emitted. When all the colors of visible light are mixed, the result is white light.

So what does this mean about our fireworks? We know that electrons and protons attract each other; thus the electrons "want" to be close to the nucleus. We will include energy levels in our model of the structure of the atom. Light is given off because when the salts are heated, the electrons are excited and move to a higher energy state. When the sample is heated, these electrons are moved farther from the nucleus, but will go back to their original state. When the electrons return to their original (ground) state, energy is released, sometimes in the form of visible light. It is quite significant that the light is not white but has a characteristic color associated with each different salt. This must mean that not all wavelengths of light are present when a metallic salt (such as copper(II) chloride) is heated. In other words, the electrons cannot go to any excited state and then return to any other state. If this occurred, all wavelengths of light would be emitted, and the light would be white. It must mean that the electrons can go to only certain energy levels. This would account for only certain colors being emitted.

This is a surprising and non-intuitive result. Energy levels in an atom are "quantized;" that is, only certain levels are possible. This means, for example, that electrons can exist at one level or another level, but not in between the levels. The first person who tried to make sense of this idea was Neils Bohr who assumed a planetary-type model in which the electrons would orbit the nucleus much like we envision the planets orbiting the sun. That is, the electrons had known, predictable pathways. It is important to realize that this model is fundamentally incorrect. It assumes we know where the electrons are, and we know where they are going. The model we accept today just is not that simple. It turns out that we simply do not know how the electron travels.

The next step in trying to make sense of atomic structure was to consider the electron as moving as a wave (like electromagnetic radiation). While the specifics of this are beyond a first-year chemistry course, realize that this has the following results: 1) waves have a probability function associated with them and 2) waves can add up and cancel each other, thus we can get regions of high and low probability (and even zero probability of finding an electron). The firefly analogy in your text (Section 11.6) is a good one to read and understand in terms of how to think about electrons in an atom. Realize that the "size" or radius of an atom is arbitrary and that we usually consider that there is a 90% probability of finding an electron in the orbital (an unfortunate term that does not mean to suggest that the electron orbits the nucleus).

Imagine the firefly analogy but with more fireflies. The spherical pattern of one firefly makes sense. But what if there were more fireflies and (like electrons) they repelled each other? What would the patterns look like? These patterns would be impossible for us to predict. These, though, are analogous to the other orbitals such as the *p* and *d* orbitals. The shapes of these are not intuitive but have been mathematically determined. It is not expected that you look at the shape of a *p* orbital, for example, and say, "Of course that's the probability region of finding the electrons!" You would have no way of predicting this exact shape. Realize though that, as we could expect, the more electrons there are in an energy level, the more complex and large the orbitals become. Realize as well that orbitals are not physical "things" but regions of probability.

The periodic table is considered in more detail in this chapter. Make sure to use it as a resource, not look at it as just another thing to memorize. The text provides an excellent discussion of this, and you should be able to answer the following:

1. How are electron configurations consistent with the placement on the table? You don't have to memorize all configurations and should be able to quickly tell the configuration from the table.
2. In Chapter 4 we saw <u>what</u> the most stable charges were for many ions. Now you should be able to explain <u>why</u> they are the most stable charges.
3. Explain (don't just state) the trends of atomic radii and ionization energy, and explain how they are related to one another.

LEARNING REVIEW

1. Which of the following represents the wavelength of electromagnetic radiation?

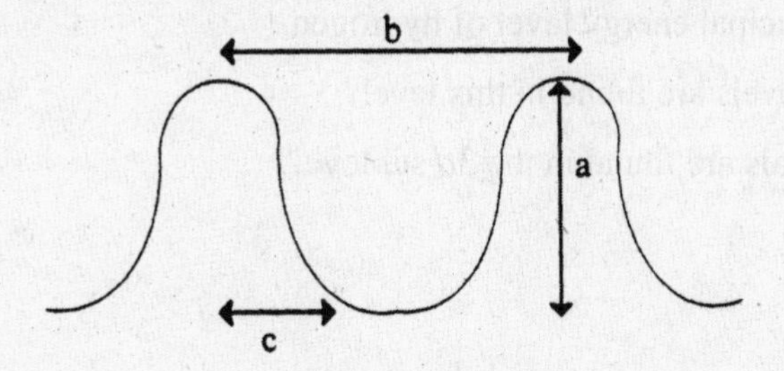

2. How does a microwave oven warm food?
3. Which has the shorter wavelength, ultraviolet light or infrared light? See Figure 11.4 in your textbook for help with electromagnetic radiation.
4. Light can be thought of as waves of energy. There is also evidence that light exists in another form. What is the other form?
5. What is meant by the terms "ground state" and "excited state" of an atom?
6. A sample of helium atoms absorbs energy. Will the photons of light emitted by the helium atoms be found at all wavelengths? Explain your answer.
7. Which energy level represents the ground state?

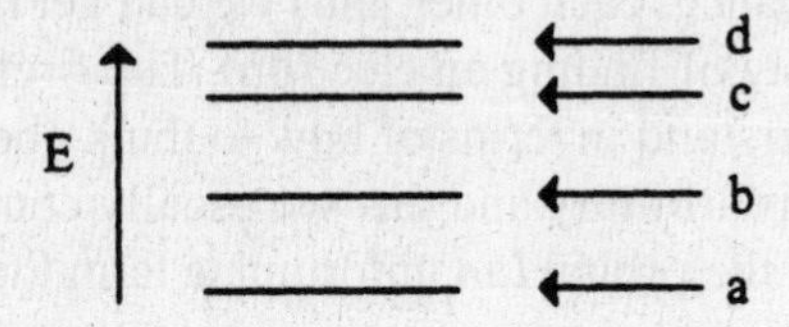

8. Which quantum has the greatest energy?

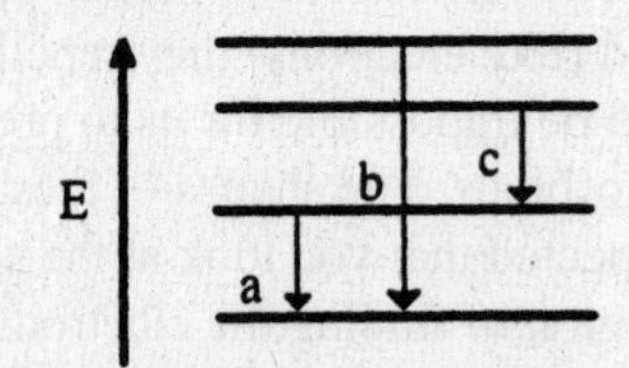

9. What does the Bohr model of the atom say about electron movement?
10. What characteristics of light led de Broglie and Schrödinger to formulate a new model of the atom?
11. Which of the following statements about the wave mechanical model of the atom are true, and which are false?
    a. The probability of finding an electron is the same in any location within an orbital.
    b. An electron will probably spend most of its time close to the nucleus.
    c. Electrons travel in circular orbits around the nucleus.
12. Which of the following statements about an orbital are true, and which are false?
    a. An electron will be found inside an orbital 90% of the time.
    b. An electron travels around the surface of an orbital.
    c. An electron cannot be found outside an orbital.
13. Consider the third principal energy level of hydrogen.
    a. How many sublevels are found in this level?
    b. How many orbitals are found in the 3*d* sublevel?

c. Which shape represents a 3*p* orbital?

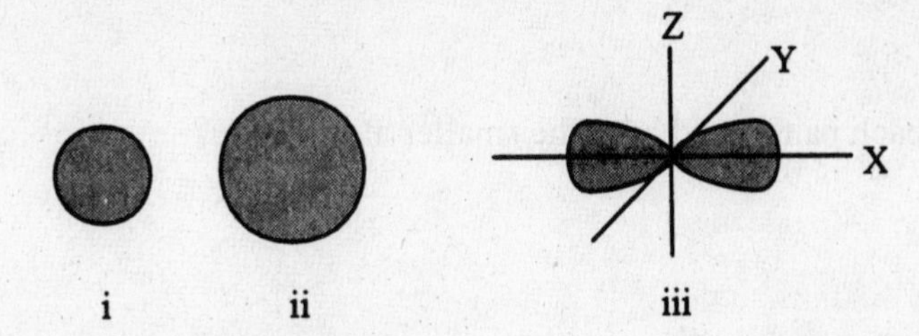

14. How many sublevels do you think would be found in the n=5 principal energy level?

15. What is meant by each part of the orbital symbol $4p_x^1$?

16. How many electrons can occupy an orbital?

17. How does each column of the periodic table relate to electron configuration?

Use a periodic table such as the one found inside the front cover of your textbook to help answer questions 18, 19, and 20.

18.
    a. Write the complete electron configuration and the complete orbital diagram for aluminum.
    b. How many valence electrons and how many core electrons does aluminum have?

19. How many valence electrons are found in the elements beryllium, magnesium, calcium and strontium?

20. Write an electron configuration and box diagram for the elements below.
    a. vanadium
    b. copper
    c. bromine
    d. tin

21. How does each row of the periodic table relate to electron configuration?

22. What is characteristic of the electron configuration of the noble gases?

23. Which orbitals are filling in the lanthanide series elements?

24. Decide whether the elements below are representative elements or transition metals.
    a. Ar
    b. Fe
    c. N
    d. Sr

25. Which element in each pair would have a lower ionization energy?
    a. F and C
    b. O and As
    c. Ca and Br

d. Li and Rb

e. Ne and Rn

f. Sr and Br

26. Which element in each pair would have the smaller atomic size?

a. Ne and Xe

b. In and I

c. Na and Cs

d. F and Sr

e. Ba and Bi

f. Cl and Al

ANSWERS TO LEARNING REVIEW

1. Wavelength is the distance from crest to crest or trough to trough, so the correct answer is b.

2. The kind of electromagnetic radiation generated by the oven, microwave radiation, is of the right frequency to be absorbed by water in food. As the water molecules in food absorb microwave energy, their movement increases. The extra energy is transferred to other molecules in the food when they collide with the water molecules, so heat is transferred from rapidly moving water molecules to other molecules and causes the food to heat up.

3. Infrared light has a wavelength of 10^{-4} meters and ultraviolet light has a wavelength of around 10^{-8} meters, so ultraviolet light has a shorter wavelength than infrared light.

4. There is evidence that light consists of packets of energy called photons.

5. The lowest possible energy state of an atom is the ground state. When an atom has absorbed excess energy it is in an excited state.

6. When helium atoms absorb energy, some of the electrons move from the ground state to an excited state. When helium atoms lose this excess energy they will often emit light. The light is *not* of just any wavelength. Only certain wavelengths corresponding to the differences in energy, are allowed.

7. The energy level marked "a" is the ground state because it is the level with the lowest amount of energy.

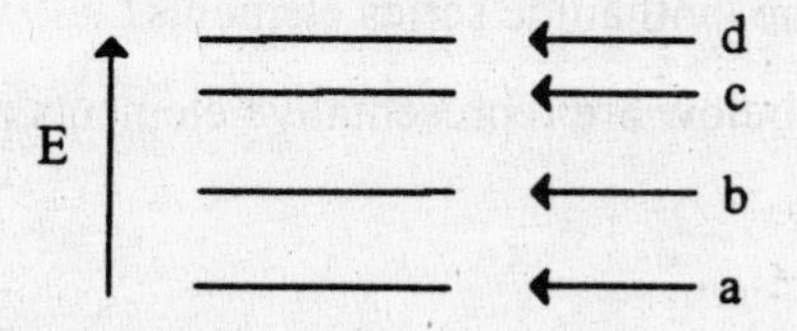

8. The quantum marked by b has the highest amount of energy because this excited state has more energy than the excited states below it.

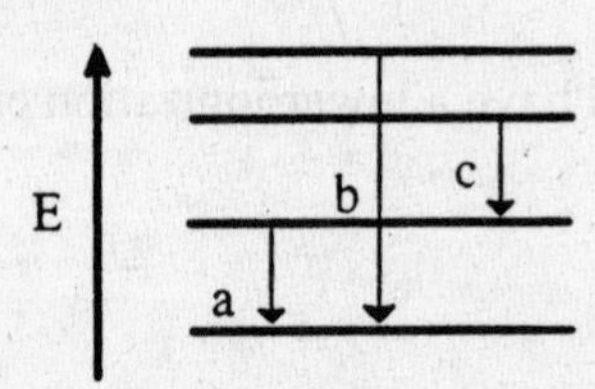

9. Bohr's model said that electrons move in circular orbits around the nucleus. Each circular orbit represents an excited state farther from the nucleus.

10. The fact that light could act both as a wave and as a particle led de Broglie and Schrödinger to suggest that an electron might also exhibit characteristics of both waves and particles. When Schrödinger used these ideas to analyze the problem mathematically, the wave mechanical model of the atom was the result.

11.

   a. The probability of finding an electron within an orbital is 90%, but some locations within the orbital shape are more likely to contain an electron at any given time than are others. So this statement is false.

   b. The electron does tend to spend most of its time around the nucleus. This statement is true.

   c. Bohr thought that electrons traveled in circular orbits around the nucleus, but the current model says that electrons are found in orbitals. We do not know their exact paths. This statement is false.

12.

   a. The orbital shape represents a probability cloud. It is true that an electron will be found within the orbital 90% of the time.

   b. The orbital marks the area of 90% probability. It does not mark the surface on which the electron travels. This statement is false.

   c. Ten percent of the time, an electron will be found outside the orbital. This statement is false.

13.

   a. The third principal energy level, n=3, contains three sublevels.

   b. There are five orbitals in the 3*d* sublevel.

   c. i and ii represent *s* orbitals; iii is a *p* orbital.

14. Each principal energy level has n sublevels, so the fifth principal energy level, n=5, would have five sublevels.

15. 4 is the principal energy level, *p* is the sublevel type, x is the specific orbital within the *p* sublevel, and the superscript 1 says that there is one electron in the orbital.

16. Each orbital can hold two electrons.

17. The period or row number indicates which *s* and *p* orbitals are being filled for elements in that row. For example, antimony, which is in row five, has filled its 5*s* orbitals and 4*d* orbitals and is filling 5*p* orbitals.

18.

   a. $1s^2 2s^2 2p^6 3s^2 3p^1$

1*s*	2*s*	2*p*			3*s*	3*p*		
↑↓	↑↓	↑↓	↑↓	↑↓	↑↓	↑		

   b. Aluminum has three valence electrons and ten core electrons.

19. Each of these elements has two valence electrons.

20.

a. 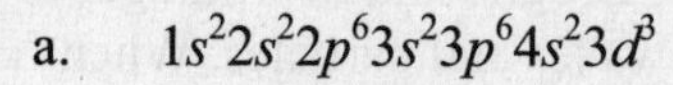
$1s^22s^22p^63s^23p^64s^23d^3$

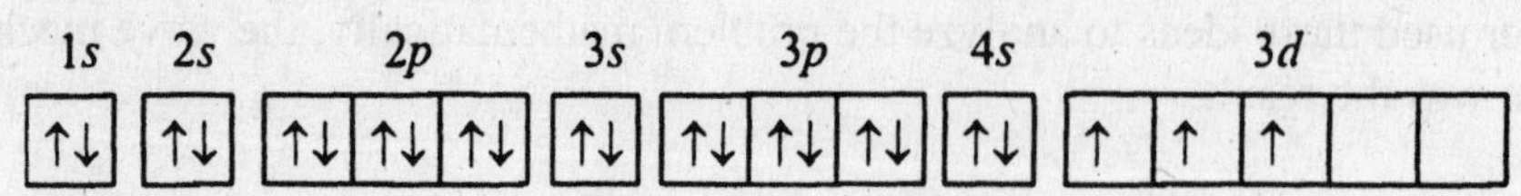

b. $1s^22s^22p^63s^23p^64s^13d^{10}$ or $[Ar]4s^13d^{10}$

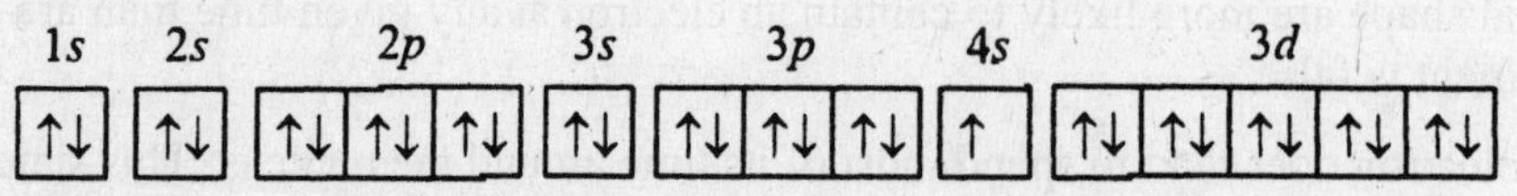

Note that copper does not completely fill 4*s* before filling 3*d*.

c. $1s^22s^22p^63s^23p^64s^23d^{10}4p^5$ or $[Ar]4s^23d^{10}4p^5$

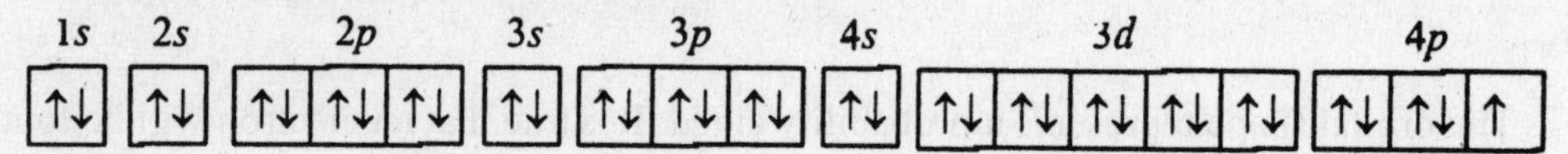

d. $1s^22s^22p^63s^23p^64s^23d^{10}4p^65s^24d^{10}5p^2$

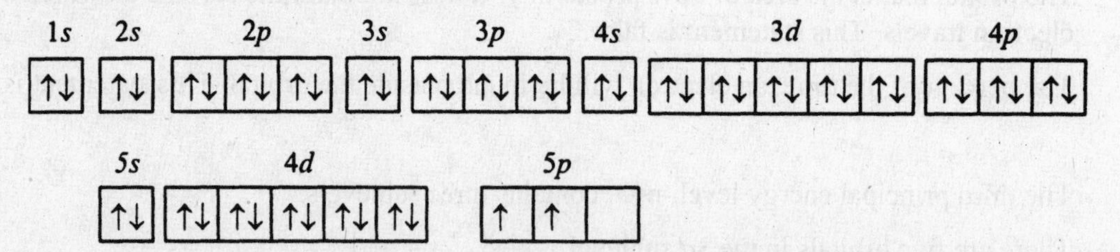

21. The group number on the top of each column of the periodic table is the same as the sum of n*s* + n*p* electrons for the highest principal energy level and is equal to the number of valence electrons for that element. For example, sulfur is in Group 6, and its electron configuration is $1s^22s^22p^63s^23p^4$. The sum n*s* + n*p* for n=3 is six, which is the same as the group number. Sulfur has six valence electrons, $3s^2$ and $3p^4$.

22. The *s* and *p* orbitals for the principal energy level, which is the same as the row number, are full. That is, all the *s* and *p* orbitals contain a maximum of two electrons for a total of eight electrons.

23. In the lanthanide series elements, the 4*f* orbitals are filling.

24.

a. Representative element (noble gas)

b. Transition metal

c. Representative element

d. Representative element (alkaline earth metal)

25.

a. Carbon would have a lower ionization energy than fluorine.

b. Arsenic would have a lower ionization energy than oxygen.

c. Calcium would have a lower ionization energy than bromine.

d. Rubidium would have a lower ionization energy than lithium.

e. Radon would have a lower ionization energy than neon.

f. Strontium would have a lower ionization energy than bromine.

26.

a. Neon would have a smaller atomic size than xenon.

b. Iodine would have a smaller atomic size than indium.

c. Sodium would have a smaller atomic size than cesium.

d. Fluorine would have a smaller atomic size than strontium.

e. Bismuth would have a smaller atomic size than barium.

f. Chlorine would have a smaller atomic size than aluminum.

CHAPTER 12

Chemical Bonding

INTRODUCTION

There are 109 elements now known, which seems like a lot of elements, yet the elements can combine to produce a far greater number of different molecules. One of the most important substances in our environment, water, is made from two hydrogen atoms and one oxygen atom that have bonded together. Understanding how and why the elements combine is a fundamental chemical concept that helps chemists predict the structures and properties of new molecules.

CHAPTER DISCUSSION

Why do atoms bond to form molecules? How do atoms bond to form molecules? These are questions that Dalton's model could not address. But after studying Chapter 11 in your text, we are ready to address them.

Let's first consider a very simple molecule, H_2. Why does H_2 exist? Why is it more stable than individual H atoms? Perhaps the simplest way to begin thinking about this is to realize that a hydrogen atom consists of an electron and a proton. Thus the electron from one hydrogen atom could be attracted to the proton of another hydrogen atom if the atoms were close enough together. There is always potential attraction between atoms because of this. The model we will use in this chapter requires that we consider the orbital theory from Chapter 11. Imagine two hydrogen atoms as represented in Figure 12.1 of your text. The first picture has the hydrogen atoms very far apart. The attraction of one atom to another (electron to proton) is negligible. But what about when these atoms are brought closer together? Electrons, which are negatively charged, will be attracted to a positive proton. Therefore, it makes sense that the electrons will spend most of the time between the two protons. This is not to say that the electrons are always there (recall from Chapter 11 that we cannot say for certain where the electrons are and how they move), but we can predict that the negatively charged electrons will be most attracted to the region between the two positively charged protons.

We want to maximize this attraction, and Figure 12.1(b) shows that the molecule is most stable when the hydrogen atoms are close together. How close will the atoms become? While the electrons are attracted to the protons, the protons repel each other. Therefore there is a limit to how close the atoms will be in the molecule.

Thus, the H_2 molecule is said to have a perfectly covalent bond. The electrons are shared between the atoms equally (each hydrogen has the same attraction for the electrons), and the electrons are lower in energy (better off) as part of the molecule than as individual atoms.

This is a simple view of covalent molecules that does not answer questions such as "Why doesn't H_3 or He_2 exist?" We can hypothesize that certain numbers of electrons will be too repulsive to allow for a stable molecule, but this model does not help us predict which molecules will be stable and which will not. We will discuss this further with Lewis structures. For now, realize that we can state that, by sharing electrons, each hydrogen in the molecule can be said to have two electrons. And by having two electrons, the outer energy level (the first energy level in this case) is complete like that in the stable noble gas helium. Because it is the electron structure that is responsible for chemical activity, the H_2 molecule should be stable. This does not mean that a filled energy level causes stability, but it appears to be related to stability. Like all models, this is a simplification, but it is a good way to first start thinking about molecules.

We have considered a molecule with two of the same atoms. What about the other end of the spectrum? That is, what about a molecule with two very different atoms? For example, what about sodium chloride (NaCl)? At first it seems reasonable to think of this molecule in the same way as we did with H_2. That is, there should be an attraction between these atoms because of the attraction between the electrons and protons. Thus we might be tempted to represent the molecule in the same way we did H_2. However, in the case of H_2, the hydrogen atoms exhibited equal attraction for the electrons. This is not true between sodium and chlorine. It turns out that chlorine has a much greater attraction for electrons than does sodium (we will consider this later). Thus, instead of the electrons being shared between the atoms, it is actually the case that ions are formed. An electron will be transferred from the sodium atom to the chlorine atom. We should represent the sodium chloride molecule like Figure 12.4(c) in your text. This is known as an ionic bond.

These are what we might call the two extreme cases. In the first case electrons were perfectly shared, and in the second case an electron was transferred from one atom to another. But what about intermediate cases? For example, consider the CO (carbon monoxide) molecule. While these atoms will not share electrons equally, the difference in their attraction for electrons is not so great that an electron will be transferred. In this case, the oxygen has more attraction for the electrons than does carbon. There is therefore a greater probability of finding an electron nearer the oxygen than the carbon. This can be represented like in Figure 12.4(b) of your text. Note that this molecule has partial positive and partial negative regions (not as extreme as ions, but not perfectly covalent either). This is termed a dipole moment, and the bond is considered polar covalent. It is covalent because electrons are shared but polar because the sharing is unequal and leads to partial charges.

It is best to think of bonding as a continuum, not as three separate types. That is, the more alike the atoms are with respect to attraction for electrons, the more covalent the bond is; the more different they are, the more ionic the bond is. This is represented in Table 12.1 of your text.

This brings about another trend called electronegativity, which is the attraction an atom in a molecule feels for a shared electron. This is discussed in detail in Section 12.2 of your text. As with atomic radii and ionization energy trends from Chapter 11, make sure to understand these trends, not merely memorize them.

Now that we have a better understanding of chemical bonds, we can go on to other questions. For example, can we predict that H_2O is a stable molecule? We know that water (H_2O) is stable, but can our model support this? And can it lead us to better understand the properties of water? The answer to both of these questions is yes.

The Localized Electron Model

At this point we would like to show how the electrons (outer shell or valence electrons) are arranged in the molecule and to describe the geometry of the molecules (which is an important function of the properties of the molecule).

To do this we will use a very simple model called the localized electron model. This model, like all models, is a simplification, but it serves to answer our questions and make correct predictions and is relatively simple to use. In this model, we assume that we know where the electrons in a molecule are located. Recall that we predict the electrons in a covalent bond will spend a great deal of time between atoms. Therefore, we assume that the electrons are fixed (localized) between the atoms. Again, these are simplifications, but they serve our purposes. Also, we assume that electrons will be shared in a molecule so that the individual atoms will have eight electrons (known as the octet rule) with the exception of hydrogen, which will have two electrons. There are some exceptions to this (as always) but these general guidelines work surprisingly well.

Sections 12.6 and 12.7 provide good guidelines for drawing Lewis structures, which show how the valence electrons are distributed in the molecule. There are also many examples of these, along with notable exceptions, so we will not belabor the point here. Becoming proficient at drawing Lewis structures requires practice. But make sure you also understand what the Lewis structures represent (as described above).

From Lewis structures we can determine the geometry of the molecules using the VSEPR model. Again, this model is quite simple but works well. The basic premise is that the electron pairs (both bonding and lone pairs) repel each other, and the geometry is a direct result of minimizing the repulsions between these electron pairs. Table 12.4 gives a nice overview of these geometries, along with example molecules.

You must draw the Lewis structure before you can determine the geometry of the molecule because the lone pairs do affect the geometry. For example, note in Table 12.4 the molecules BF_3 and NH_3. Both of these molecules seem similar from their formulas. However, they have different geometries because they have different Lewis structures.

LEARNING REVIEW

1. What are the two kinds of bonds that can form between atoms?
2. What kind of bond forms between
   a. two identical atoms?
   b. a metal and a nonmetal?
3. What is meant by a polar covalent bond?
4. Which of these choices has an ionic bond?
   a. CO
   b. $CaBr_2$
   c. HBr
   d. Cl_2
5. Arrange the following atoms based on electronegativity. Put the most electronegative atom on the right and the least electronegative atom on the left.

   P Al Cl Mg

6. Which bond is the most polar? Which bond is the least polar?
   a. P–Cl
   b. H–H
   c. N–H
   d. C–F
7. Write electron configurations for both reactants and products for the reactions below.
   a. $Mg + Cl_2 \rightarrow Mg^{2+} + 2Cl^-$
   b. $2Li + S \rightarrow 2LI^+ + S^{2-}$

8. When two nitrogen atoms combine to form a nitrogen molecule, how many electrons are shared to give each atom a complete octet?

9. How many electrons do each of the atoms below need to gain, lose or share to achieve a noble-gas valence electron configuration?

   a. S

   b. Mg

   c. C

10. In which of the atom/ion pairs is the ion *smaller* than the atom?

    a. S/S^{2-}

    b. Ca/Ca^{2+}

    c. Li/Li^{2+}

    d. I/I^{-}

11. Draw Lewis structures for these ionic compounds.

    a. MgS

    b. Na_2O

12. Write valence electron configurations for the following atoms.

    a. B

    b. Sr

    c. Kr

    d. Cl

13. Draw Lewis structures for the molecules or ions below.

    a. H_2S

    b. HCCH

    c. PO_4^{3-}

    d. HI

    e. PCl_3

14. Some molecules have Lewis structures that violate the octet rule. Draw a probable Lewis structure for BeI_2.

15. Use the VSEPR model to determine the molecular structure of each of the molecules below.

    a. SbF_3

    b. BH_3

    c. SiH_4

    d. H₂C=O (H and H bonded to C, C=O)

ANSWERS TO LEARNING REVIEW

1. When atoms combine they form either ionic bonds or covalent bonds. In an ionic bond, electrons are completely transferred. In a covalent bond, electrons are shared between atoms. Ionic bonds usually form when a metal and a nonmetal react. Covalent bonds usually form when two nonmetals react.

2.
   a. When two identical atoms bond, a nonpolar covalent bond forms.

   b. When a metal and a nonmetal bond, an ionic bond forms.

3. Electrons shared between two atoms are not always shared equally. Sometimes the electrons are attracted to one of the atoms more than the other. The atom in a bond that attracts the electron pair will have an extra electron part of the time and so bear a partial negative charge. The atom that does not strongly attract the electron pair it shares will be electron-deficient part of the time and so will bear a positive charge. The kind of covalent bond where the electrons are not shared equally is called a polar covalent bond.

4. Ionic bonds are formed when a metal loses an electron to a nonmetal. Among these choices, the only bond between a metal and a nonmetal is the bond formed between calcium and bromine to form calcium bromide. So the correct answer is b.

5. Elements on the right side of the periodic table have higher electronegativity values than do elements on the left side of the periodic table. The atom with the highest electronegativity would be Cl because it is in the upper right-hand corner of the periodic table; then P, then Al, and Mg, on the left side of the periodic table, is the lowest.

   Mg Al P Cl

6. To determine the polarity of a bond, subtract the electronegativity of the least electronegative atom from the electronegativity of the most electronegative atom. The largest difference is the most polar bond, and the smallest difference is the least polar bond.

   a. P is 2.1 while Cl is 3.0. The difference is 0.9.

   b. Both H atoms are 2.1. The difference is 0.0.

   c. N is 3.0 while H is 2.1. The difference is 0.9.

   d. C is 2.5 while F is 4.0. The difference is 1.5.

   The most polar of these bonds is the C–F bond because the difference in electronegativities is the highest, 1.5. The least polar bond is that which is formed between two hydrogen atoms. The difference in electronegativities is zero, which means that this bond is completely nonpolar.

7.
   a. $Mg + Cl_2 \rightarrow Mg^{2+} + 2Cl^-$

   $1s^22s^22p^63s^2 + 1s^22s^22p^63s^23p^5 \rightarrow 1s^22s^22p^6 + 2(1s^22s^22p^63s^23p^6)$

   b. $2Li + S \rightarrow 2Li^+ + S^{2-}$

   $2(1s^22s^1) + 1s^22s^22p^63s^23p^4 \rightarrow 2(1s^2) + 1s^22s^22p^63s^23p^6$

8. Nitrogen is in Group 5 of the periodic table, which means that it has five valence electrons. Most elements obey the octet rule when they bond with other atoms, so a nitrogen molecule should be composed of two nitrogen atoms, each with a complete octet. When two nitrogen atoms bond, they can only achieve an octet by sharing six electrons, three from each nitrogen atom. So in a nitrogen molecule each nitrogen atom has two valence electrons of its own and shares six with another nitrogen atom.

9.

a. Sulfur is in Group 6 of the periodic table, so it has six valence electrons. Sulfur needs two more electrons to fill its valence shell. Sulfur often gains two electrons to become a sulfide ion, S^{2-}. The sulfide ion has the same electron configuration as argon.

b. Magnesium is in Group 2 of the periodic table and has two valence electrons. Magnesium atoms lose two electrons to form the Mg^{2+} ion. The magnesium ion has the same electron configuration as neon.

c. Carbon is in Group 4 of the periodic table, so it has four valence electrons. Each carbon atom will share four valence electrons with other atoms. When a carbon atom shares four electrons with other atoms, it has the same electron configuration as neon.

10.

a. The sulfide ion has gained two electrons, so it is larger than the sulfur atom.

b. The calcium ion has lost two electrons, so it is smaller than the calcium atom.

c. The lithium ion has lost one electron, so it is smaller than the lithium atom.

d. The iodide ion has gained one electron, so it is larger than the iodine atom.

11.

a. In ionic compounds, the electrons lost from the metal are transferred to the nonmetal so that each atom can achieve a noble-gas configuration.

$$Mg^{2+} \qquad :\ddot{\underset{..}{S}}:^{2-}$$

b. Two sodium atoms each lose one electron to an oxygen atom so that all three atoms achieve a noble-gas configuration.

$$Na^{+} \quad :\ddot{\underset{..}{O}}:^{2-} \quad Na^{+}$$

12.

a. B $2s^1 2p^1$

b. Sr $5s^2$

c. Kr $4s^2 4p^6$

d. Cl $3s^2 3p^5$

13.

a. Step 1. Find the total number of valence electrons in all the atoms. There are eight valence electrons in a molecule of H_2S. Six come from sulfur and one each from hydrogen.

Step 2. Begin distributing the available valence electrons by putting a pair of electrons between each atom. An electron pair is often symbolized with a line.

H—S—H

Step 3. We have used up four of the available electrons. There are four more to distribute. Each of the two hydrogen atoms is satisfied, so the four extra electrons must appear as unshared pairs on the sulfur atom.

H—S̤̈—H

Now all atoms satisfy either the octet or the duet rule.

b. Step 1. The total number of valence electrons for two carbon atoms and two hydrogen atoms is ten, four from each carbon atom and one each from the hydrogen atoms.

Step 2. Arrange the electron pairs between atoms.

H—C—C—H

Step 3. There are now four electrons left. If we distribute two around each carbon atom we have

H—C̈—C̈—H

The carbon atoms do not fulfill the octet rule, and we have run out of electrons. Some of the atoms must share more than one pair of electrons in order for the octet rule to be satisfied. We cannot share more electrons between the carbon and the hydrogen atoms because hydrogen already satisfies the duet rule. So the electrons must be shared between the two carbon atoms. Let's begin by sharing one more pair of electrons. The result is

H—C=C—H

This leaves us with two electrons left.

H—Ċ=Ċ—H

Carbon is still not satisfied. Let's share another electron pair between the two carbon atoms.

H—C≡C—H

Now all of the valence electrons are used, and both hydrogen and carbon are satisfied.

c. Step 1. The phosphate ion, PO_4^{3-}, has a total of 32 valence electrons: 24 from the oxygen atoms, five from phosphorus, and three extra electrons because of the 3^- charge on the phosphate ion.

Step 2. Distribute pairs of electrons among the atoms.

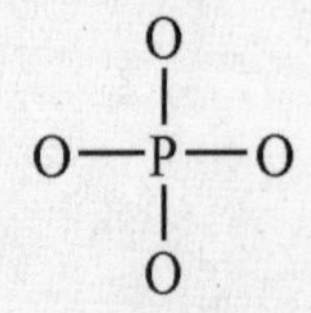

Step 3. There are 24 electrons left. Begin arranging the electrons around the atoms to satisfy the octet rule for all the atoms. Three unshared pairs of electrons surround each oxygen atom. The complete Lewis structure is

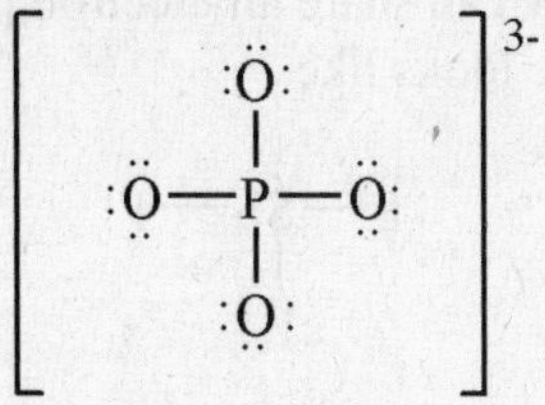

```
        ..       3-
     [ :O: ]
        |
   ..   |   ..
  :O—P—O:
   ..   |   ..
        |
       :O:
        ..
```

d. Step 1. HI has a total of eight valence electrons.

Step 2. Arrange an electron pair between the two atoms.

H—I

Step 3. Now, hydrogen is satisfied, and there are six electrons left. Arrange the remaining electron pairs around the iodine atom.

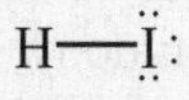

```
   ..
H—I:
   ..
```

e. Step 1. Phosphorus trichloride has a total of 26 valence electrons.

Step 2. Arrange electron pairs among the atoms.

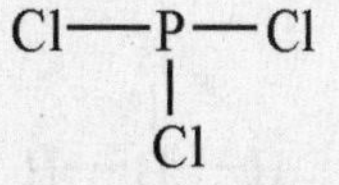

```
Cl—P—Cl
   |
   Cl
```

Step 3. There are 20 electrons left. Then arrange the electrons around each atom until each has eight electrons. Three unshared pairs surround each chlorine atom, and the phosphorus atom has one unshared pair.

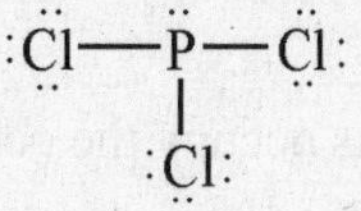

```
 ..   ..   ..
:Cl—P—Cl:
 ..   |   ..
     :Cl:
      ..
```

14. Step 1. Add together the valence electrons contributed by each atom in the compound. Each iodide has seven, and beryllium has two, for a total of 16 valence electrons.

Step 2. Arrange the electrons in pairs between the atoms.

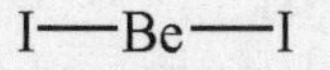

I—Be—I

Step 3. This leaves 12 electrons to distribute as unshared pairs. The iodide atoms obey the octet rule, but beryllium atoms often do not. Beryllium is electron deficient and will have only four valence electrons.

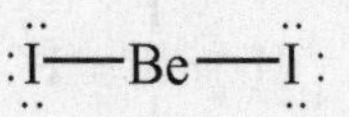

```
 ..       ..
:I—Be—I:
 ..       ..
```

15.

a. Step 1. Draw the Lewis structure for the molecule. Antimony has five valence electrons, and fluorine has seven. Antimony can share an electron pair with each of the three fluorine atoms, so the Lewis structure looks like

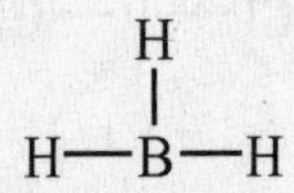

Step 2. Count the electron pairs, and arrange them so they are as far apart as possible. When three pairs of electrons are shared and there is one pair left that is not shared, we can arrange the electrons in a tetrahedron.

Step 3. Determine the positions of the atoms using the electron pairs as a guide. The fluorine atoms will occupy the corners of a tetrahedron. The lone pair of electrons on the antimony atom will occupy the fourth corner of the tetrahedron.

Step 4. Determine the molecular structure using the positions of the atoms. In this molecule there are three atoms surrounding antimony, and each one is located in the corner of a tetrahedron. Because only three corners of the tetrahedron are occupied by atoms, this molecular shape is a trigonal pyramid.

b. Step 1. The Lewis structure of boron trihydride is

H
|
H—B—H

Boron has three valence electrons and shares electrons with three hydrogen atoms. Note that boron is an exception to the octet rule.

Step 2. The three pairs of electrons spread themselves out as far as possible to form a trigonal planar shape.

Step 3. The three hydrogen atoms occupy the corners of a triangle.

Step 4. This molecular structure is called trigonal planar. The bond angle is 120°.

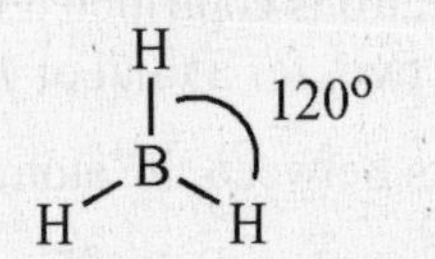

c. Step 1. Silicon has four valence electrons, and each of the hydrogen atoms has one. Silicon can share electron pairs with each of the four hydrogen atoms so the Lewis structure looks like

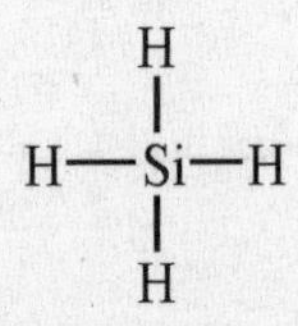

Step 2. When four pairs of electrons are shared, the VSEPR model predicts that they will spread apart to form a tetrahedron.

Step 3. There are four hydrogen atoms bonded to a central silicon atom. The four hydrogen atoms will occupy the four corners of a tetrahedron.

Step 4. The molecular structure of this molecule will be a tetrahedron. The bond angle is 109.5°.

H
| 109.5°
Si
H H H

d. Step 1. The Lewis structure for this molecule shows that there is a carbon-oxygen double bond.

O
||
H—C—H

Step 2. The three effective pairs of electrons spread themselves out as far as possible to form a trigonal planar shape.

Step 3. The two hydrogen atoms and the oxygen atom occupy the corners of a triangle.

Step 4. This molecular structure is trigonal planar. The bond angle around the center atom is 120°.

O 120°
||
C
H H

CHAPTER 13

Gases

INTRODUCTION

We live in a gaseous environment. The air we breathe is a mixture of gases. Yet gases are not as conspicuous as liquids and solids are, and it is easy to overlook their significance. Understanding why gases behave as they do can help us understand everyday occurrences such as "low pressure systems" in our weather and the apparent decrease of the amount of air in car tires in cold weather. This chapter focuses on how gases behave under various circumstances and why they behave as they do.

CHAPTER DISCUSSION

One of the best byproducts of studying gas laws is achieving a greater understanding of the concept of models and getting a better feel for scientific thinking. As with most scientific endeavors, one of the goals is to be able to explain real phenomena; for example, explaining how a hot-air balloon works or why our ears "pop" when flying in an airplane. Observations are made, and when they appear to be consistent, they are termed laws (for example, Boyle's law). It is important to understand that these laws tell us "what" (for example, the pressure of a gas is inversely proportional to its volume if the amount of gas and temperature are held constant), but do not tell us "why." Models (or theories) are developed to explain the "whys." In the case of gases, the kinetic molecular theory is offered as an explanation. Remember, though, that models are not reality; that is, they include simplifications.

Sections 13.2, 13.3, and 13.4 give us observations that, for the most part, consist of facts. For example, place a balloon filled with air in the freezer, and it will shrink. Thus temperature and the volume of a gas are directly related as long as pressure and the amount of gas remain constant (as the temperature decreases, the volume decreases). These observations should not be surprising to you, but they are presented in these sections as mathematical expressions.

There is no need to memorize all of these relationships because we bring them all together in the ideal gas equation:

$$PV = nRT$$

This equation brings together all of the gas laws discussed in Chapter 13. There are quite a few example problems in Sections 13.1-13.5. After reading these sections and working on the examples, you should be able to show how all the other gas laws (Boyle's law, etc.) can be derived from the ideal gas law equation. Pay attention to what is constant and why. This way you need to know only this one equation.

For example, consider Charles's law. This law states that the volume of a gas is directly dependent on the temperature of the gas (in Kelvin) provided that the amount of gas and the pressure are held constant. Thus V and T are changing, and P and n are constant (as is R, of course, as always). Let's rearrange the ideal gas law equation so that the factors that change are on one side, and the factors that are constant are on the other:

$$\frac{V}{T} = \frac{nR}{P} = \text{constant}$$

Because R, P, and n are all constant, the factor "$\frac{nR}{P}$" must be constant as well. Therefore, we can write this as

$$\frac{V}{T} = \frac{nR}{P} = \text{constant}$$

or

$$\frac{V}{T} = \text{constant}$$

This shows us that the ratio "$\frac{V}{T}$" is a constant, as long as P and n are held constant (and the temperature is in Kelvin). We can rewrite this as

$$\frac{V_1}{T_1} = \frac{V_2}{T_2}$$

which is the mathematical representation of Charles's law you will find in Section 13.3 of your text. You should spend some time and derive the other laws using this same approach.

These relationships explain what happens, but not why. To explain why, we need to consider a model of gases. Section 13.6 in your text discusses Dalton's law of partial pressures. Read through this section and understand that this law is an observation; that is, it is a fact. The text also says that this law tells us two important things about gases that are important enough to repeat here:

1. The volume of the individual gas particle must not be very important.
2. The forces among the particles must not be very important.

Make sure to understand how Dalton's law of partial pressures tells us this. Discuss this with friends or an instructor. These conclusions lead us to our model, the kinetic molecular theory (KMT).

The KMT is discussed in Section 13.8. Notice that two of the premises come from the above conclusions, which come from observations. This is how model development works. That is, we make observations and use the significance of these observations in developing a model. For example, consider two of the assumptions in the KMT:

1. The volume (size) of the individual particles can be assumed to be zero.
2. The particles are assumed not to attract or to repel each other.

Note how these two premises are actually "ideal" cases that come from Dalton's law of partial pressures. That is, Dalton's law of partial pressures leads us to believe that the volume of gas particles is not important, nor are the forces among these particles. In the KMT, we assume that not only are these factors not important, they are non-existent. Why do we do this? Because it is easier; that is, it makes for a simpler model. And recall, our goal is to make the model as simple as we can as long as it explains the observations we want it to explain.

You should strive to understand the significance of the KMT. Read through sections 13.8 and 13.9 and make sure you can explain the relationships among pressure, volume, amount of gas, and temperature.

LEARNING REVIEW

1. Gas in compressed gas cylinders is usually under a great deal of pressure. If the gas in a particular cylinder has a pressure of 2500 psi, how many torr is this?

2. Convert each of the units of pressure below.

   a. 0.408 atm to torr

   b. 68,471 Pa to mm Hg

   c. 50.9 psi to atm

3. The relationship observed by Boyle between volume and pressure is

   a. linear

   b. proportional

   c. inversely proportional

   d. no relationship

4. Examine the cylinder with a moveable piston. If the piston moves downward, causing the volume of the gas to decrease, will the pressure of the gas become larger or smaller?

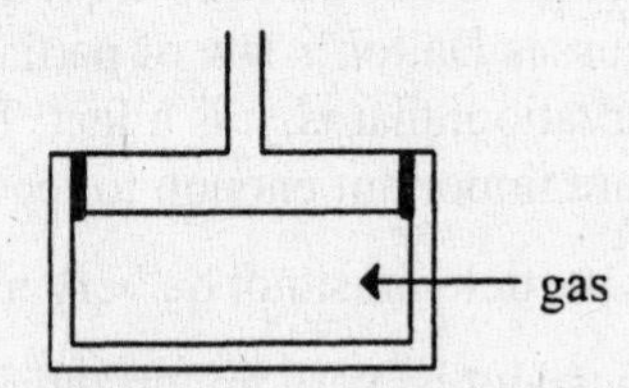

5. A sample of nitrogen gas at 2.4 atm has a volume of 50.3 L. If the pressure is decreased to 1.9 atm, will the new volume be greater or smaller? What is the new volume?

6. Absolute zero is the temperature at which gases have zero volume. But zero volume has never been measured in the laboratory. So how do we know at what temperature the volume of a gas equals zero? Be as specific as possible.

7. Which of the following equations cannot be derived from Charles's law, $V = bT$?

   a. $\frac{V}{T} = b$

   b. $\frac{V_1}{T_1} = \frac{V_2}{T_2}$

   c. $\frac{V_1}{T_1} + \frac{V_2}{T_2} = b$

   d. $\frac{3V}{3T} = b$

8. A container with a moveable piston contains 0.89 L methane gas at 100.5 °C. If the temperature of the gas rises by 11.3 °C, what is the new volume of the gas?

9. Avogadro's law describes the relationship between the amount (number of moles) of a gas and the volume of the gas. Under what conditions of temperature and pressure is Avogadro's law true?

10. Where does the universal gas constant, *R*, come from?

11. Use the ideal gas law to solve the problems below.

    a. A sample of chlorine gas at 543 torr has a volume of 21.6 L. If the temperature of the chlorine is 0 °C, how many moles of chlorine gas are present?

    b. Poisonous carbon monoxide gas is a product of the internal combustion engine. If 1.2 mol of CO at 11°C and 102 mm Hg are present in a container, what will be the volume of the CO gas?

    c. 0.45 mol of a gas has a pressure of 299 torr at 300 °C and a volume of 53.8 L. At the same temperature and pressure, the volume of the gas is decreased to 39.7 L. How many moles of gas are present after the volume has changed?

12. Which statements about Dalton's law of partial pressure are true and which are false?

    a. The total pressure (P_{TOTAL}) of a mixture of gases is independent of the sizes of the gas particles.

    b. Attractive forces between gas particles are important in determining P_{TOTAL}.

    c. For ideal gases, P_{TOTAL} depends solely on the total number of moles of gas for any temperature and volume.

13. Assume that a sample of humid air contains only nitrogen gas, oxygen gas and water vapor. If the atmospheric pressure is 745 mm Hg and the partial pressure of N_2 is 566 mm Hg and of oxygen is 140. mm Hg, what is the partial pressure of water vapor in the air?

14. A 7.5-L mixture of gases is produced by mixing 4.0 L of N_2 at 450 torr, 3.5 L of O_2 at 252 torr, and 0.21 L of CO_2 at 150 torr. If the temperature is held constant at 65°C, what is the total pressure of the mixture?

15. Explain the difference between a law and a model.

16. Which statements about the Kinetic Molecular Theory are true, and which are false?

    a. The postulates of the Kinetic Molecular Theory are true for all gases at all temperatures and pressures.

    b. Gas particles are assumed to either attract or repel each other.

    c. The distance between individual gas particles is much greater than the volume of an individual gas particle.

    d. As the Kelvin temperature of a gas increases, the average kinetic energy increases.

17. According to the Kinetic Molecular Theory, what are we measuring when we measure the temperature of a gas?

18. Carbon dioxide is produced during the combustion of liquid propane fuel.

$$C_3H_8 + 5O_2 \rightarrow 3CO_2 + 4H_2O$$

    If 5.0 kg of propane are burned at 1.000 atm pressure and 400.°C, what volume of CO_2 gas is produced?

19. A sample of fluorine gas has a volume of 19.9 L at STP. How many moles of fluorine are in the sample?

ANSWERS TO LEARNING REVIEW

1. There is no conversion factor directly between psi and torr. However, we can convert psi to atm and atm to torr.

$$2500 \text{ psi} \times \frac{1.000 \text{ atm}}{14.69 \text{ psi}} \times \frac{760.00 \text{ torr}}{1.000 \text{ atm}} = 1.3 \times 10^5 \text{ torr}$$

2.

a. $0.408 \text{ atm} \times \dfrac{760.0 \text{ torr}}{1.000 \text{ atm}} = 310. \text{ torr}$

b. $68471 \text{ Pa} \times \dfrac{1.000 \text{ atm}}{101{,}325 \text{ Pa}} \times \dfrac{760.0 \text{ mm Hg}}{1.000 \text{ atm}} = 513.6 \text{ mm Hg}$

c. $50.9 \text{ psi} \times \dfrac{1.000 \text{ atm}}{14.69 \text{ psi}} = 3.46 \text{ atm}$

3. Boyle observed that as the volume increased, the pressure decreased. Pressure and volume are inversely proportional. The correct answer is c.

4. If the volume of gas inside the cylinder becomes smaller, then the pressure of the gas will become larger.

5. This problem provides pressure and volume data. Boyle's law relates pressure to volume. We can use $P_1V_1 = P_2V_2$, which is one way of writing Boyle's law, to solve this problem. In this equation, P_1 and V_1 represent initial, or starting, conditions. P_2 and V_2 represent final or changed conditions. 2.4 atm is the initial pressure, P_1, and 50.3 L is the initial volume, V_1. Pressure has changed so P_2 is 1.9 atm, and we are asked for the new volume, V_2. First, will the new volume be larger or smaller? The pressure decreases from 2.4 atm to 1.9 atm. There is an inverse relationship between temperature and pressure, so if the pressure decreases, we would expect the volume to increase.

Rearrange the equation to isolate V_2 on one side.

$$P_1V_1 = P_2V_2$$

$$V_1 \times \frac{P_1}{P_2} = V_2 \times \frac{P_2}{P_2}$$

$$V_2 = V_1 \times \frac{P_1}{P_2}$$

Now substitute values into the equation.

$$V_2 = 50.3 \text{ L} \times \frac{2.4 \text{ atm}}{1.9 \text{ atm}}$$

$$V_2 = 64 \text{ L } N_2$$

This answer makes sense. The pressure has decreased from 2.4 atm to 1.9 atm, so we would expect the volume to increase. The volume has in fact increased from 50.3 L to 64 L. There is an inverse relationship between pressure and volume.

6. We can measure the volume of a gas at various temperatures, some of which are very cold but not quite absolute zero. We can then plot on a graph each temperature and volume pair.

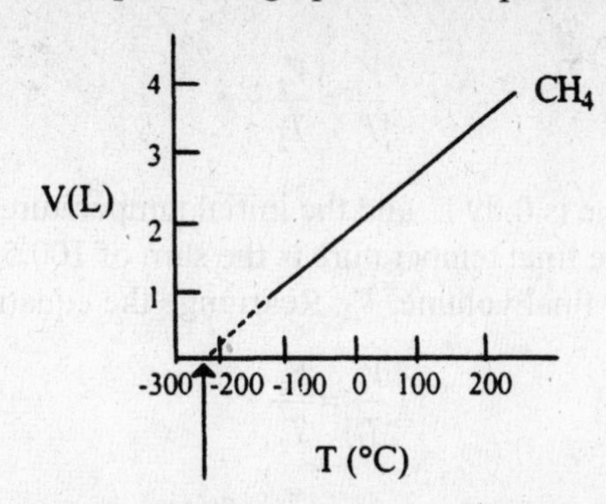

Although we cannot measure the volume of a gas at −273.2 °C, we can determine at what temperature the volume would be zero by drawing a continuation of the straight line of the graph and noting where the temperature axis hits the line. This has been done for many different gases, and the result is always the same; the temperature at which the volume of a gas would be zero is −273.2°C.

7.

a. Charles's law can be written as $V = bT$, where V is volume, T is temperature and b is a constant. We can rearrange the equation by dividing both sides by T.

$$\frac{V}{T} = b \times \frac{T}{T}$$

$$b = \frac{V}{T}$$

So, $\frac{V}{T} = b$ can be derived from Charles's law.

b. If $\frac{V_1}{T_1} = b$ and $\frac{V_2}{T_2} = b$, then it is also true that $\frac{V_1}{T_1} = \frac{V_2}{T_2}$, so this relationship can also be derived from $V = bT$.

c. We know that $\frac{V_1}{T_1} = b$ and that $\frac{V_2}{T_2} = b$. We also know that $\frac{V_1}{T_1} = \frac{V_2}{T_2}$, but we cannot add $\frac{V_1}{T_1}$ and $\frac{V_2}{T_2}$ together to equal the constant b. So this relationship cannot be derived from $V = bT$.

d. By multiplying temperature and volume by the same value, we are not changing the equation, because $\frac{3V}{3T}$ and $\frac{V}{T}$ are both equal to the constant b.

$$\frac{3V}{3T} = b = \frac{V}{T}$$

so this relationship can also be derived from $V = \frac{b}{T}$.

8. This problem provides temperature and volume data and asks about the volume of methane gas when the temperature changes. We can use Charles's law to solve the problem. One form of Charles's law is

$$\frac{V_1}{T_1} = \frac{V_2}{T_2}$$

The initial volume of methane is 0.89 L, and the initial temperature is 100.5 °C. The rise in temperature is 11.3 °C, so the final temperature is the sum of 100.5 °C plus 11.3 °C, or 111.8 °C. The unknown quantity is the final volume, V_2. Rearrange the equation to isolate V_2 on one side.

$$\frac{V_1}{T_1} = \frac{V_2}{T_2}$$

$$V_1 \times \frac{T_2}{T_1} = V_2 \times \frac{T_2}{T_2}$$

$$V_2 = V_1 \times \frac{T_2}{T_1}$$

Because both the initial and final temperatures are given in °C, we must convert to Kelvin.

initial temperature (T_1)

$T_K = T_{°C} + 273$

$T_K = 100.5 + 273$

$T_K = 374$

final temperature (T_2)

$T_K = 111.8 + 273$

$T_K = 385$

Now substitute values into the equation.

$$V_2 = 0.89 \text{ L methane} \times \frac{385 \text{ K}}{374 \text{ K}}$$

$$V_2 = 0.92 \text{ L methane}$$

This answer makes sense. The temperature has increased, so the volume should also increase. There is a direct relationship between the temperature in Kelvin of a gas and its volume.

9. Avogadro's law, which expresses a relationship between the volume of a gas and the number of moles of that gas, is true only when the gas temperature and pressure are constant.

10. When the three laws that relate volume to pressure, to temperature and to the number of moles are combined, the constants that are present in each individual law are combined into one constant, called R, the universal gas constant. When volume is expressed in liters, pressure in atmospheres, and temperature in Kelvin, R has units of L·atm/K·mol and a value of 0.08206 L atm/K mol.

11. The ideal gas law can be expressed as $PV = nRT$. Remember that temperature must be expressed in kelvins, pressure in atmospheres, and volume in liters. If any of the quantities given in the problem are expressed in other units, we must convert them before we can use the ideal gas law.

a. This problem provides the pressure of chlorine gas in torr and the temperature in °C. Before we proceed let's convert torr to atmosphere and °C to K.

$$543 \text{ torr} \times \frac{1.000 \text{ atm}}{760.0 \text{ torr}} = 0.714 \text{ atm}$$

$$T_K = T_{°C} + 273$$

$$T_K = 0 + 273$$

$$T_K = 273$$

We are provided with pressure, volume, and temperature data, and we are asked for the number of moles. Rearrange the ideal gas law equation to isolate n on one side.

$$PV = nRT$$

$$\frac{PV}{RT} = n \times \frac{RT}{RT}$$

$$n = \frac{PV}{RT}$$

Now substitute values into the equation.

$$n = \frac{(0.714 \text{ atm})(21.6 \text{ L } Cl_2)}{(0.08206 \text{ L atm/K mol})(273 \text{ K})}$$

$$n = 0.688 \text{ mol } Cl_2$$

b. This problem asks us to calculate the volume of carbon monoxide gas, CO, given temperature, pressure and the number of moles of gas. The pressure of CO is given in mm Hg and temperature in °C. Before we proceed, we need to convert mm Hg to atmospheres and °C to K.

$$102 \text{ mm Hg} \times \frac{1.000 \text{ atm}}{760.0 \text{ mm Hg}} = 0.134 \text{ atm}$$

$$T_K = T_{°C} + 273$$

$$T_K = 11 + 273$$

$$T_K = 284$$

We can use the ideal gas law to solve this problem because we are provided with temperature, pressure and the number of moles of gas. Rearrange the ideal gas law to isolate V on one side.

$$PV = nRT$$

$$V \times \frac{P}{P} = \frac{nRT}{P}$$

$$V = \frac{nRT}{P}$$

Now substitute values into the equation.

$$V = \frac{(1.2 \text{ mol})(0.08206 \text{ L atm/mol K})(284 \text{ K})}{(0.134 \text{ atm})}$$

$$V = 210 \text{ L CO}$$

c. In this problem we are provided with information about two different gases. The information about the first gas is complete. That is, we know the number of moles, the temperature, the volume and the pressure of the gas. We know the temperature, the volume and pressure of the second gas and are asked to calculate the number of moles, n. We could use Avogadro's law, which relates volume and moles, to solve this problem, but we are asked to use the ideal gas law. Because we know the temperature and the volume and the pressure of the second gas, we can use the ideal gas law to calculate the number of moles. First convert pressure, which is given in torr, to atm and temperature, which is given in °C, to K.

$$299 \text{ torr} \times \frac{1.000 \text{ atm}}{760.0 \text{ torr}} = 0.393 \text{ atm}$$

$$T_K = T_{°C} + 273$$

$$T_K = 300. + 273$$

$$T_K = 573$$

Rearrange the ideal gas law to isolate n on one side.

$$PV = nRT$$

$$\frac{PV}{RT} = n \times \frac{RT}{RT}$$

$$n = \frac{PV}{RT}$$

Now substitute values into the equation.

$$n = \frac{(0.393 \text{ atm})(39.7 \text{ L})}{(0.08206 \text{ L atm/mol K})(573 \text{ K})}$$

$$n = 0.332 \text{ mol}$$

12.

a. The total pressure (P_{TOTAL}) depends on only the total quantity of gas, not on the kind and size of the particles. So this statement is true.

b. Because P_{TOTAL} does not depend on the identity of the gas particles but only on the quantity of particles, this statement is false.

c. The pressure of a single gas or a mixture of gases depends on the number of moles, and also the temperature and volume. A mixture of gases follows the ideal gas law, just as a single gas does. So this statement is false.

13. P_{TOTAL} represents the pressure exerted by the gases in the atmosphere, which in this case is equal to 745 mm Hg. P_{TOTAL} equals $P_{OXYGEN} + P_{NITROGEN} + P_{WATER\ VAPOR}$. The two major components of air are nitrogen and oxygen. If $P_{NITROGEN}$ equals 566 mm Hg, and P_{OXYGEN} equals 140. mm Hg, then we can use the following equation to calculate $P_{WATER\ VAPOR}$.

$$P_{TOTAL} = P_{OXYGEN} + P_{NITROGEN} + P_{WATER\ VAPOR}$$

Rearrange the equation to isolate $P_{WATER\ VAPOR}$ on one side.

$$P_{WATER\ VAPOR} = P_{TOTAL} - (P_{OXYGEN} + P_{NITROGEN})$$

Now substitute values into the equation.

$$P_{WATER\ VAPOR} = 745\text{ mm Hg} - (140.\text{ mm Hg} + 566\text{ Hg})$$

$$P_{WATER\ VAPOR} = 39\text{ mm Hg.}$$

14. We are asked to calculate P_{TOTAL} for a mixture of three gases. To determine P_{TOTAL}, we need to know the partial pressure of each individual gas. Calculating the partial pressure of any gas by using the ideal gas law requires us to know the number of moles, *n*, of that gas. So we need to find a way to calculate *n*. We are given the initial volume of each gas, the initial pressure, and the initial temperature, so we can use the ideal gas law in the form $n = \frac{PV}{RT}$ to find the number of moles. Because the number of moles of each gas does not change when the gases are mixed together, we can use the value of *n* we calculated for each gas before they are mixed together to calculate the partial pressure of each gas under the conditions present when the gases are mixed. Because pressure is given in torr and temperature in °C, convert to atmospheres and Kelvin before proceeding.

The temperature is held constant at 65 °C. Convert to atmospheres and Kelvin before proceeding.

$$T_K = T_{°C} + 273$$

$$T_K = 65 + 273$$

$$T_K = 338$$

For nitrogen

$$450\text{ torr} \times \frac{1.000\text{ atm}}{760.0\text{ torr}} = 0.59\text{ atm}$$

$$n_{N_2} = \frac{(0.59\text{ atm})(4.0\text{ L }N_2)}{(0.08206\text{ L atm/mol K})(338\text{ K})} = 0.085\text{ mol }N_2$$

For oxygen

$$252\text{ torr} \times \frac{1.000\text{ atm}}{760.0\text{ torr}} = 0.332\text{ atm}$$

$$n_{O_2} = \frac{(0.332\text{ atm})(3.5\text{ L }O_2)}{(0.08206\text{ L atm/mol K})(338\text{ K})} = 0.042\text{ mol }O_2$$

For carbon dioxide

$$150 \text{ torr} \times \frac{1.000 \text{ atm}}{760.0 \text{ torr}} = 0.197 \text{ atm}$$

$$n_{CO_2} = \frac{(0.197 \text{ atm})(0.21 \text{ L } CO_2)}{(0.08206 \text{ L atm/mol K})(338 \text{ K})} = 1.5 \times 10^{-3} \text{ mol } CO_2$$

So now we know the number of moles of each gas. We can use the ideal gas law to calculate the partial pressure of each gas. We will need to use the volume of the gas **mixture**, not the volume of each individual gas. Use the ideal gas law with P isolated on one side.

$$P = \frac{nRT}{V}$$

$$P_{N_2} = \frac{(0.085 \text{ mol } N_2)(0.08206 \text{ L atm/mol K})(338 \text{ K})}{7.5 \text{ L}} = 0.31 \text{ atm}$$

$$P_{O_2} = \frac{(0.042 \text{ mol } O_2)(0.08206 \text{ L atm/mol K})(338 \text{ K})}{7.5 \text{ L}} = 0.16 \text{ atm}$$

$$P_{CO_2} = \frac{(1.5 \times 10^{-3} \text{ mol } CO_2)(0.08206 \text{ L atm/mol K})(338 \text{ K})}{7.5 \text{ L}} = 5.5 \times 10^{-3} \text{ atm}$$

The total pressure, P_{TOTAL}, can be calculated from

$$P_{TOTAL} = 0.31 \text{ atm} + 0.16 \text{ atm} + 5.5 \times 10^{-3} \text{ atm}$$

$$P_{TOTAL} = 0.48 \text{ atm}$$

15. A law is a generalization about behavior that has been drawn from repeated observation about nature. A model is a theory that attempts to explain why a certain behavior is observed.

16.

   a. False. Gases obey the postulates of the Kinetic Molecular Theory best at high temperatures and/or high pressures. When these conditions are not met, gases can exhibit other kinds of behavior.

   b. False. In postulate four of the Kinetic Molecular Theory it is assumed that gas particles neither attract nor repel each other.

   c. True. In postulate two of the Kinetic Molecular Theory, it is assumed that individual gas particles are very small compared to the distances between the particles.

   d. True. The kinetic energy of a gas is proportional to the kelvin temperature.

17. When we measure the temperature of a gas, we are actually measuring how fast the gas particles are moving. Postulate five of the Kinetic Molecular Theory results from the observation that as the temperature of a gas increases, the movement of gas particles increases.

18. This problem provides the pressure, the temperature and the quantity in kg of a sample of propane gas. We are asked to find the volume of another gas, carbon dioxide, produced when propane burns. We know some information about carbon dioxide. We know that the pressure is 1 atm and the temperature is 400°C. If we knew the number of moles of carbon dioxide, we could use the ideal gas law to calculate the volume of carbon dioxide.

$$PV = nRT$$

? ?

We know the mass of propane that reacts to form CO_2, and we have a balanced equation, so we can calculate the moles of carbon dioxide using the mole ratio from the balanced equation. But first let's calculate the number of moles of propane that are equivalent to 5.0 kg of propane.

$$5.0 \text{ kg } C_3H_8 \times \frac{1000 \text{ g } C_3H_8}{1 \text{ kg } C_3H_8} \times \frac{1 \text{ mol } C_3H_8}{44.09 \text{ g } C_3H_8} = 110 \text{ mol } C_3H_8$$

Now that we know the number of moles of propane, we can use the mole ratio to calculate the moles of CO_2 that would be produced. The balanced equation tells us that every mole of C_3H_8 produces 3 moles of CO_2 so the correct mole ratio is $\frac{3 \text{ mol } CO_2}{1 \text{ mol } C_3H_8}$

$$110 \text{ mol } C_3H_8 \times \frac{3 \text{ mol } CO_2}{1 \text{ mol } C_3H_8} = 330 \text{ mol } CO_2$$

Now we know the number of moles of CO_2, the pressure and the temperature, so we can use the ideal gas law to calculate the volume. First isolate V on one side.

$$PV = nRT$$

$$\frac{P}{P} \times V = \frac{nRT}{P}$$

$$V = \frac{nRT}{P}$$

Because the temperature is given in °C, we must convert to Kelvin.

$$T_K = T_{°C} + 273$$

$$T_K = 400. + 273$$

$$T_K = 673 \text{ K}$$

Now substitute values into the equation.

$$V = \frac{(330 \text{ mol } C_3H_8)(0.08206 \text{ L atm/mol K})(673 \text{ K})}{(1.0 \text{ atm})}$$

$$V = 1.8 \times 10^4 \text{ L } CO_2$$

19. We can solve this problem two ways. In the first method we can use the ideal gas law because we know the volume, the temperature and the pressure. Rearrange the equation to isolate moles on one side.

$$PV = nRT$$

$$\frac{PV}{RT} = n \times \frac{RT}{RT}$$

$$n = \frac{PV}{RT}$$

The fluorine gas is at STP, which means that the temperature is 0 °C, and the pressure is 1 atm. We need to convert °C to Kelvin before we proceed.

$$T_K = 0 + 273$$

$$T_K = 273$$

Now substitute values into the equation.

$$n_{F_2} = \frac{(1.0 \text{ atm})(19.9 \text{ L})}{(0.08206 \text{ L atm/mol K})(273 \text{ K})}$$

$$n_{F_2} = 0.888 \text{ mol}$$

In the second method we can use the conversion 1 mol of an ideal gas = 22.4 liters of gas. This conversion works only if the gas is at STP. In this problem we are told that the temperature and pressure are at STP.

$$19.9 \text{ L } F_2 \times \frac{1.00 \text{ mol } F_2}{22.4 \text{ L } F_2} = 0.888 \text{ mol } F_2$$

Both methods gave the same answer, but we can use only the second method when the temperature is 0 °C (273 K) and the pressure is 1.00 atmosphere.

CHAPTER 14

Liquids and Solids

INTRODUCTION

This chapter discusses the properties of liquids and solids. You will learn what makes the particles in solids stay together and why some liquids boil at higher temperatures than others. One liquid of particular importance is water. Liquid water is one of the most important parts of our environment, and we routinely use water to dissolve many kinds of substances. As we will see, water has some unusual properties.

CHAPTER DISCUSSION

Why is it that, when we constantly apply heat to ice, it eventually melts to form liquid water? And why does this water, when heated, eventually form steam?

Also, how can it be that different chemicals are different phases at the same temperature? For example, sodium chloride is a solid, water is a liquid, and nitrogen is a gas, all at normal room conditions.

These are the types of questions that you should be able to answer by studying Chapter 14 in your text.

The bonds that we discussed in Chapter 12 were intramolecular bonds. That is, these were forces that held atoms together in a molecule. We now want to consider intermolecular bonds, or bonds that occur among molecules to form a solid or a liquid. Figure 14.1 provides a microscopic representation of a solid, a liquid, and a gas.

Recall from our discussion of the VSEPR model (Chapter 12) that water is a bent or v-shaped molecule, and is therefore polar. In other words, there are regions of partial positive and negative charges in a water molecule, and these opposite charges provide the force that holds these molecules together (see Figures 14.4, 14.5, and 14.6 in your text). This attraction is termed dipole-dipole attraction. For molecules such as water, which has a hydrogen bonded to oxygen that is very electronegative (see Chapter 12), these dipole-dipole attractions are particularly strong and termed hydrogen bonding. Hydrogen bonding accounts for some of the quite unusual properties of water. For example, because these intermolecular bonds are rather strong, water is a liquid at room conditions (water is a rather small and “lightweight” molecule, and its boiling point is quite high for its molar mass and size).

So why does ice turn to water which turns to steam as we heat our sample? We know that temperature is a measure of kinetic energy (see Chapter 10), and an increase in kinetic energy will cause the molecules to move more quickly. The more movement, the more disorder. Look again at Figure 14.1, and note that solids are more ordered than liquids, which are more ordered than gases.

So this provides us a way of thinking about phase changes. But what about the fact that some chemicals are solids, some are liquids, and some are gases at the same room conditions? This too can be answered with our model of intermolecular forces.

Consider, for example, nitrogen gas (N_2). Draw the Lewis structure for N_2, and convince yourself that it is a nonpolar molecule. Because of this there should be no permanent dipole-dipole interactions. Because there seems to be no strong force holding these molecules together, it should make sense that N_2 is a gas. In fact, it is not obvious that one N_2 molecule would have any attraction for another N_2 molecule.

However, we know that N_2 can be a liquid if we get the temperature low enough (you may have seen or heard of liquid nitrogen before). Therefore there must be intermolecular attractions of some sort among all molecules. These forces are termed London dispersion forces and come about from a random uneven distribution of electrons around the nucleus (see Figure 14.8). As the temperature is lowered, the molecules slow down. If they are slowed down enough (which is why liquid nitrogen is very cold) the molecules have enough attraction to come together to form a liquid.

Look at Figure 14.8, and think about this question, "Will a molecule with more electrons have a greater or smaller chance of having an uneven distribution of electrons?" Think about this before reading on.

The more electrons there are, the greater the chance that there will be an uneven distribution of electrons. The more electrons, the greater number of ways there are to be unevenly distributed. As a simple example, consider the following. If there were only two electrons in a diatomic molecule, the electrons could be evenly distributed or both could be on one side of the molecule. But with ten electrons there are many more ways for uneven distribution (six on one side, four on the other; seven and three, etc.) and many combinations of these. This should lead us to believe that the more electrons a molecule has, the stronger the London dispersion forces will be (the greater the uneven distribution, the stronger the attraction).

Consider, for example, I_2, Br_2, and Cl_2. Each of these is nonpolar and therefore relies on only London dispersion forces for intermolecular attractions. Thus, the type of attraction for each of these is the same. But I_2 has the most electrons, and Cl_2 has the least. We would predict, then, that I_2 molecules would exhibit the most attraction for one another, and Cl_2 would exhibit the least. As it turns out, I_2 is a solid, Br_2 is a liquid, and Cl_2 is a gas, all at room conditions. While our simple model could not predict that I_2 should be a solid, it is pleasing to note that we can predict the order of attraction.

You should be able to answer the following questions after studying Chapter 14:

1. Why is sodium chloride a solid with such a high melting point?
2. Why is water a liquid but methane (CH_4) a gas at room conditions?

LEARNING REVIEW

1. Why does ice float on the surface of liquid water?
2. Which of the following statements about water are true, and which are false?
    a. At 0 °C water can be either a solid or a liquid.
    b. The normal boiling point of water is 97.5 °C at 1 atm pressure.
    c. Forces among water molecules are called intramolecular forces.
3. Why does it take more energy to convert liquid water to steam than it does to convert ice to liquid water?
4. How much energy is required when 15.0 g of liquid water at 75 °C is heated to 100.°C, and then converted to steam at 100.°C? The molar heat of vaporization for liquid water is 40.6 kJ/mol, and the specific heat capacity of liquid water is 4.18 J/g °C.
5. How much energy is required to convert 6.0 g liquid water to steam at 100°C?
6. Which bond is stronger, a covalent bond or a hydrogen bond?
7. Draw a diagram showing how three NH_3 molecules can form hydrogen bonds.
8. Why does water, H_2O, boil at a higher temperature than H_2S?
9. Explain how London dispersion forces form between two molecules of argon.

10. Match the following terms with the correct definition.

Terms	Definitions
a. condensation	the pressure exerted by a liquid at equilibrium with its vapor
b. vapor pressure	a balance between two opposite processes
c. vaporization	vapor molecules form a liquid
d. equilibrium	a liquid becomes a gas

11. Which of the molecules in each pair would have the greater vapor pressure? To help you determine which of these molecules is polar, review Sections 12.2 and 12.3 of your textbook.

    a. CH_4, CH_3OH

    b. H_2O, H_2S

12. Name the three types of crystalline solids, and give an example of each.

13. What intermolecular force is responsible for holding each of the following compounds together in the solid state?

    a. KCl

    b. HF

    c. $SiCl_4$

14. Mixtures of metallic elements form alloys. What are the two kinds of alloys, and how do they differ from each other?

ANSWERS TO LEARNING REVIEW

1. Ice is less dense than the water, so it floats on the surface. The density of ice (solid water) is less than the density of liquid water because water expands as it freezes. This means that ice has the same mass as water but occupies a greater volume. Using the formula density equals mass/volume, as the volume becomes greater, the density is less.

2.

    a. 0°C is the normal freezing point of water. Both liquids and solids can exist at the same time at the normal freezing point of a liquid. This statement is true.

    b. The normal boiling point of water is 100°C at 1 atm pressure, not 97.5°C. This statement is false.

    c. Forces among water molecules are called **inter**molecular forces. Forces among atoms of a single water molecule are called **intra**molecular forces. This statement is false.

3. Converting ice to liquid water involves overcoming relatively few intermolecular forces. This is true because the molecules in both solid and liquid water are close together, so there is relatively little disruption when the change from solid to liquid occurs. However, going from liquid water to steam requires overcoming the intermolecular forces among water molecules, so a great amount of energy is required. The water molecules in steam are very far apart compared to water molecules in the liquid state.

4. In this problem, there are two changes occurring. Each change must be considered separately when calculating the overall amount of energy needed. In the first change, we heat the liquid water from 75°C to 100.°C. This conversion requires the use of the specific heat capacity for water along with the equation below.

Energy required = specific heat capacity × mass × temperature change

$$Q = c \times m \times \Delta T$$

We are given the specific heat capacity for liquid water and the mass of water, and we can calculate the change in temperature. If the final temperature is 100.°C and the beginning temperature is 75 °C, then the change in temperature, ΔT, must be 25 °C.

Now substitute into the equation.

$$Q = c \times m \times \Delta T$$

$$Q = 4.18 \text{ J/g °C} \times 15.0 \text{ g water} \times 25 \text{ °C}$$

$$Q = 1600 \text{ J}$$

In the second change, liquid water at 100 °C is converted to steam at 100.°C. We are going from the liquid state to a gas, but the temperature remains 100.°C. To calculate the amount of energy needed during this change, we can use the molar heat of vaporization for water as a conversion factor.

molar heat of vaporization = 40.6 kJ/mol H_2O

This conversion factor requires us to know the number of moles of steam. The problem tells us we have 15.0 g steam. We can calculate the number of moles from the molar mass of water..

$$15.0 \text{ g } H_2O \times \frac{1 \text{ mol } H_2O}{18.02 \text{ g } H_2O} = 0.832 \text{ mol } H_2O$$

Now use the molar heat of vaporization as a conversion.

$$0.832 \text{ mol } H_2O \times \frac{40.6 \text{ kJ}}{1 \text{ mol } H_2O} = 33.8 \text{ kJ}$$

To convert 15.0 g liquid water from 75 °C to 100.°C requires 1600 J, and to convert 15.0 g liquid water at 100.°C to a gas requires 33.8 kJ. To determine the total amount of energy required, we can add together the two quantities of energy. However, the units do not match. One quantity is given in Joules and the other in kiloJoules. Let's convert J to kJ.

$$1600 \text{ J} \times \frac{1 \text{ kJ}}{1000 \text{ J}} = 1.6 \text{ kJ}$$

The total amount of energy required is 1.6 kJ plus 33.8 kJ, which is equal to 35.4 kJ.

5. This problem requires only one step. We want to know how much energy is needed to convert water at 100°C to steam at 100°C. We can use the molar heat of vaporization of water as a conversion factor after we convert grams of water to moles of water.

$$6.0 \text{ g water} \times \frac{1 \text{ mol water}}{18.02 \text{ g water}} = 0.33 \text{ mol water}$$

Now use the molar heat of vaporization to convert mol water to kiloJoules.

$$0.33 \text{ mol water} \times \frac{40.6 \text{ kJ}}{1 \text{ mol water}} = 13 \text{ kJ}$$

The amount of energy required to convert 6.0 g water at 100 °C to steam at 100 °C is 13 kJ.

6. Hydrogen bonds are very strong dipole-dipole forces that occur between molecules that contain a hydrogen atom bonded to either a nitrogen atom, an oxygen atom, or a fluorine atom. Even relatively strong hydrogen bonds are only 1% as strong as covalent bonds. Covalent bonds are much stronger than hydrogen bonds.

7. Ammonia molecules can form hydrogen bonds because there is a hydrogen atom bonded to a strongly electronegative nitrogen atom, producing a strong dipole.

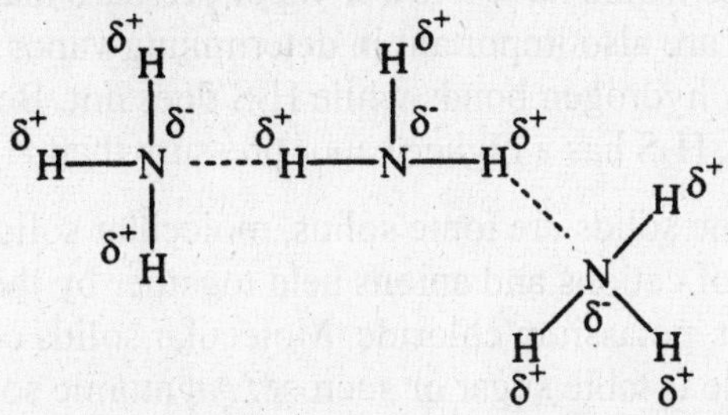

8. Both H_2O and H_2S contain two hydrogen atoms. The difference between the two molecules is the central atom, oxygen or sulfur. Oxygen is an electronegative atom while sulfur is not very electronegative. Water molecules can form strong hydrogen bonds because the O–H bond is very polar. The difference in electronegativity between oxygen and hydrogen is large. A great deal of heat energy is required to break apart the hydrogen bonds between water molecules so that the molecules can enter the vapor phase. The S–H bond in H_2S is not very polar because the difference in electronegativity between hydrogen and sulfur is small. No hydrogen bonds form between H_2S molecules, so less heat energy is required to boil liquid H_2S.

9. The weak intermolecular forces are called London dispersion forces. They are weaker than hydrogen bonds or dipole-dipole interactions. We normally assume that the electrons in atoms such as argon are distributed somewhat evenly around the atom. However, we cannot predict the path an electron will take as it moves around the atom. Sometimes more electrons are momentarily found on one side of an argon atom. This causes a small and temporary dipole. This temporary dipole can induce a dipole on an atom that may be nearby. The negative side of one argon atom is attracted to the positive side of another argon atom. These attractive forces between atoms are weak, and they do not last long. As the electrons move, the force dissipates.

10.

a. condensation: vapor molecules form a liquid

b. vapor pressure: the pressure exerted by a vapor in equilibrium with its liquid

c. vaporization: a liquid becomes a gas

d. equilibrium: a balance between two opposite processes

11. Two factors determine the relative vapor pressure of any two liquids; the molecular weights of the two molecules and the intermolecular forces.

    a. CH_3OH has a higher molecular weight, 32.0 g/mol, than CH_4, which has a molecular weight of 16.0 g/mol. In a molecule of CH_3OH, a hydrogen atom is bonded to an electronegative oxygen atom. When hydrogen atoms are bonded to electronegative atoms such as oxygen, hydrogen bonds can form among molecules. CH_3OH molecules are attracted to each other by hydrogen bonds. CH_4 molecules contain only carbon-hydrogen bonds. Carbon is not as electronegative as atoms such as oxygen, so molecules of CH_4 are not very polar. There are no hydrogen bonds among CH_4 molecules. The intermolecular forces will be weak London dispersion forces. So the vapor pressure of CH_4 will be higher than that of CH_3OH.

    b. H_2S has a higher molecular weight, 34.0 g/mol, than H_2O, 18.0 g/mol. Purely on the basis of molecular weight, H_2S would have a lower vapor pressure than water. However, intermolecular forces are also important in determining vapor pressure. Water molecules form relatively strong hydrogen bonds while H_2S does not. Because of the lack of strong intermolecular forces, H_2S has a higher vapor pressure than H_2O.

12. The three types of crystalline solids are ionic solids, molecular solids, and atomic solids. An ionic solid exists as a collection of cations and anions held together by the attractive forces between the ions. An example is the salt, potassium chloride. Molecular solids consist of molecules. There are no ions present. An example is table sugar or sucrose. An atomic solid is made from individual atoms, all the same. Pure copper metal is an example of an atomic solid.

13.

    a. KCl consists of potassium ions and chloride ions. The ions are held together in solid KCl by the forces that exist between the oppositely charged ions.

    b. When hydrogen is bonded to electronegative fluorine atoms to produce molecules of hydrogen fluoride, hydrogen bonds can form among molecules. The fluorine atom bears a partial negative charge while the hydrogen atom bears a partial positive charge.

    c. $SiCl_4$ contains only silicon-chlorine bonds. This molecule will not form intermolecular hydrogen bonds because hydrogen bonding occurs only when hydrogen bonds to electronegative atoms such as oxygen, nitrogen or fluorine. Also, this molecule is not polar because it is a balanced molecule. There are no partially positive or negative ends. So there are no hydrogen bonds and no dipole-dipole attractions among $SiCl_4$ molecules, but London dispersion forces cause weak attractive forces among the molecules.

14. An alloy is a substance that contains a mixture of elements and has metallic properties. There are two basic kinds of alloys. Substitutional alloys have some of the metal atoms replaced with other metal atoms that are approximately the same size. Interstitial alloys have some of the small spaces between metal atoms filled with atoms smaller than the metal atoms.

CHAPTER 15

Solutions

INTRODUCTION

It is important to know how much solid is dissolved in a liquid. Telling someone that you added a little sugar to their iced tea does not give them much of a clue about how sweet the tea will be. A little sugar to you might mean a lot of sugar to someone else. In this chapter you will learn ways to express concentration so that you know just how much dissolved solid is present in a given volume of liquid.

CHAPTER DISCUSSION

Many common ionic and nonionic substances dissolve in water. When an ionic substance dissolves, it breaks apart into ions. For example, when potassium sulfate, K_2SO_4, dissolves, it forms K^+ and SO_4^{2-} ions. To understand why a crystal of potassium sulfate breaks apart into ions, we need to consider the nature of the water molecule. Water molecules are polar; that is, one end of the molecule has a partial positive charge, and the other end has a partial negative charge. The positive potassium ion is attracted to the partial negative charge on the water molecule, and the negative sulfate ion is attracted to the partial positive charge on the water molecule. The K_2SO_4 crystal is pulled apart by the polarity of the water molecule. Ions in solution are surrounded by oppositely charged ends of water molecules.

Nonionic compounds such as ethyl alcohol, C_2H_5OH, also dissolve in water. The O–H of ethyl alcohol is polar, just as the O–H on the water molecule is. This means that alcohol molecules have a negative end and a positive end just as water molecules do. Alcohol molecules are attracted by water molecules and are dissolved in them.

Molecules that are not water soluble do not have positive or negative ends to be attracted to water.

One way of describing the composition of a solution is mass percent. The mass percent of a solution is the mass of solute dissolved by the total mass of the solute plus solvent, multiplied by 100%.

$$\text{mass percent} = \left(\frac{\text{mass of solute}}{\text{mass of solution}}\right) \times 100\%$$

However, the mass percent of a solution is inconvenient to use when the solvent is liquid. It is much more convenient to measure the volume of a liquid than it is to measure its mass. The most often used indication of the amount of solute in a given volume of solution is molarity. Molarity is equal to the number of moles of solute per volume of solution.

$$\text{Molarity} = \left(\frac{\text{moles of solute}}{\text{liters of solution}}\right)$$

This is often abbreviated as $M = \text{mol/L}$.

When the molarity of a solution is calculated, it is assumed that the solute is in the form it would have been in before it dissolved. One mole of Na_2SO_4, when added to water, produces two moles of Na^+ ions and one mole of SO_4^{2-} ions.

$$1.0 \text{ mol } Na_2SO_4 \rightarrow 2.0 \text{ mol } Na^+ + 1.0 \text{ mol } SO_4^{2-}$$

Compounds that do not form ions have the same molar concentration before and after they dissolve in water. Sucrose, table sugar, dissolves in water but does not form ions. 1 mol of sucrose in 1 liter of solution produces a 1 *M* solution of sucrose.

Dilution is the process of adding more solvent to a solution. When we prepare a dilute solution, we are measuring a quality of a stock solution (relatively highly concentrated solution) and adding it to water. The amount of solute in the measured portion of stock solution is the same as the amount of solute in the more dilute solution. The only thing which has changed is the volume of the solution. We have decreased the concentration of the solution by increasing the volume, but the amount of solute stays the same.

In Chapter 9, you learned to solve stoichiometry problems. From the balanced equation you could answer questions about the quantity of reactant required or the quantity of product produced. The same principles are used here, but the reactions occur in solution. In Chapter 9 we used molar mass to convert mass to moles. Here we use molarity to convert volume of solution to moles. Use the steps in Section 15.6 of your text.

LEARNING REVIEW

1. 150 mL of ethyl alcohol is mixed with 1 L of water. Which is the solute, ethyl alcohol or water?

2. Which of the molecules below would you predict to be soluble in water?

   a. CH_2-OH / $CH-OH$ / CH_2-OH (each carbon bonded in a chain)

   b. $CH_3-CH_2-CH_3$

   c. K_2SO_4

3. Three solutions are prepared by mixing the quantities of sodium chloride given below in a volume of 500 mL of solution. Which solution is the most concentrated?

   a. 55 g NaCl in 500 mL solution

   b. 127 g NaCl in 500 mL solution

   c. 105 g NaCl in 500 mL solution

4. A student stirred 5.0 g of table sugar into 250. g of hot coffee. What is the mass percent of sugar in the coffee?

5. Calculate mass percents for the solutions below.

   a. 6.5 g KOH in 250. g water

   b. 0.40 g baking soda in 2000.0 g flour

   c. 150. g acetone in 438 g water

6. A solution of HCl in water is 0.15 *M*. How many mol/L of HCl are present?

7. 150.5 g of NaOH are dissolved in water. The final volume of the solution is 3.8 L. What is the molarity of the solution?

8. Calculate the molarity of each of the solutions below.

   a. 0.62 g $AgNO_3$ in a final volume of 1.5 L solution

b. 10.6 g NaCl in a final volume of 286 mL solution

c. 152 g $Ca(NO_3)_2$ in a final volume of 0.92 L solution

9. 2.5 L of a solution of KI in water has a concentration of 0.15 *M*. How many grams of KI are in the solution?

10. What is the concentration of each ion in the following solutions?

a. 2.0 *M* H_2SO_4

b. 0.6 *M* Na_3PO_4

c. 1.5 *M* $AlCl_3$

11. How many moles of KCl are present in 1.5 L of a 0.48 *M* solution of KCl in water?

12. How many grams of NaOH are needed to make 1.50 L of a 0.650 *M* NaOH solution?

13. How many grams of K_2SO_4 are needed to make 250 mL of a 0.150 *M* K_2SO_4 solution?

14. Sodium fluoride is added to many water supplies to prevent tooth decay. How many grams of NaF must be added to a water supply so that 2.0×10^6 L of water contain 3.0×10^{-6} *M* NaF?

15. What volume of 12 *M* HCl solution is needed to make 2.5 L of 1.0 *M* HCl?

16. What volume of 18 *M* H_2SO_4 stock solution is needed to make 1855 mL of 0.65 *M* H_2SO_4?

17. If 2.5 L of solution contains 0.10 *M* $CaCl_2$, how many grams of Na_3PO_4 are needed to exactly precipitate all the calcium as $Ca_3(PO_4)_2$?

18. How many grams of $Fe(OH)_3$ can be produced by the addition of 0.25 moles of $FeCl_3$ to 1.2 L of a 0.85 *M* NaOH solution?

19. What volume of 0.15 *M* NaOH solution will react completely with 150 mL of 0.25 *M* HCl?

20. For each of the strong acids and strong bases below, give the number of equivalents present in 1 mole and the equivalent weight of each.

Material	Molar Mass	Equivalents	Equivalent Weight
1.0 mol HCl	36.46		
1.0 mol H_2SO_4	98.08		
1.0 mol KOH	56.11		

21. If 5.6 grams of phosphoric acid, H_3PO_4, are added to water so that the final volume is 125 mL, what is the normality of the solution?

ANSWERS TO LEARNING REVIEW

1. Ethyl alcohol is the solute. Water is the solvent because it is present in the largest amount.

2.

a. This molecule contains three polar O–H bonds. Each of these O–H bonds can form hydrogen bonds with water. This molecule is soluble in water.

b. This molecule has no polar bonds. There is no part of the molecule that will interact with polar water molecules. This molecule is not soluble in water.

c. Potassium sulfate is an ionic compound. Many ionic compounds dissolve in water because the charged ions are pulled from the crystal by the polar water molecules. This molecule is soluble in water.

3. Solution b, 127 g NaCl per 500 mL solution, is the most concentrated because the amount of solute per amount of solution is the greatest.

4. The mass percent of a solution can be calculated by

$$\text{mass percent} = \left(\frac{\text{mass of solute}}{\text{mass of solution}}\right) \times 100\%$$

The mass of solute is 5.0 g sugar, and the mass of the solution is the mass of solute plus the mass of the solvent or 5.0 g sugar plus 250. g coffee equals 255 g solution. The mass percent sugar is

$$\frac{5.0 \text{ g sugar}}{255 \text{ g solution}} \times 100\% = 2.0\% \text{ sugar}$$

5.

a. The mass of solute is 6.5 g KOH, and the mass of solution is 6.5 g KOH plus 250. g water, which is 257 g. The mass percent is

$$\frac{\text{mass of solute}}{\text{mass of solution}} \times 100\% = \frac{6.5 \text{ g KOH}}{257 \text{ g solution}} \times 100\% = 2.5\% \text{ KOH}$$

b. The mass of solute is 0.40 g, and the mass of solution is 0.40 g baking soda plus 2000.0 g flour, which is equal to 2000.4 g. The mass percent is

$$\frac{0.40 \text{ g baking soda}}{2000.4 \text{ g solution}} \times 100\% = 0.02\% \text{ baking soda}$$

c. The mass of solute is 150. g acetone, and the mass of solution is 150. g of acetone plus 438 g water, which is 588 g. The mass percent is:

$$\frac{150. \text{ g acetone}}{588 \text{ g solution}} \times 100\% = 25.5\% \text{ acetone}$$

6. The definition of molarity, *M*, is moles solute/liter solution. An HCl solution that is 0.15 *M* would contain 0.15 mol HC1/liter solution.

$$0.15 \ M \text{ HCl} = \frac{0.15 \text{ mol HCl}}{\text{L}}$$

7. The molarity of a solution is equal to the moles solute/liter solution. In this problem we have 150.5 g solute, NaOH. We do not know the number of moles. Using the molar mass for NaOH, we can calculate the number of moles.

$$150.5 \text{ g NaOH} \times \frac{1 \text{ mol NaOH}}{40.00 \text{ g NaOH}} = 3.763 \text{ mol NaOH}$$

Now, we can find the molarity.

$$M = \frac{\text{moles solute}}{\text{liters solution}}$$

$$M = \frac{3.763 \text{ mol NaOH}}{3.8 \text{ L}} = 0.99 \ M \text{ NaOH}$$

8.

a. First calculate the moles of $AgNO_3$.

$$0.62 \text{ g } AgNO_3 \times \frac{1 \text{ mol } AgNO_3}{169.91 \text{ g } AgNO_3} = 0.0036 \text{ mol } AgNO_3$$

Now calculate the molarity.

$$M = \frac{0.0036 \text{ mol } AgNO_3}{1.5 \text{ L}}$$

$$M = 2.4 \times 10^{-3}\ M\ AgNO_3$$

b. First calculate the moles of NaCl.

$$10.6 \text{ g NaCl} \times \frac{1 \text{ mol NaCl}}{58.44 \text{ g NaCl}} = 0.181 \text{ mol NaCl}$$

The volume of the solution is given in milliliters. We need to know the number of liters.

$$286 \text{ mL solution} \times \frac{1 \text{ L solution}}{1000 \text{ mL solution}} = 0.286 \text{ L solution}$$

Now calculate the molarity.

$$M = \frac{0.181 \text{ mol NaCl}}{0.286 \text{ L solution}}$$

$$M = 0.633\ M \text{ NaCl}$$

c. First calculate the moles of $Ca(NO_3)_2$

$$152 \text{ g } Ca(NO_3)_2 \times \frac{1 \text{ mol } Ca(NO_3)_2}{164.10 \text{ g } Ca(NO_3)_2} = 0.926 \text{ mol } Ca(NO_3)_2$$

Now calculate the molarity.

$$M = \frac{0.926 \text{ mol } Ca(NO_3)_2}{0.92 \text{ L}}$$

$$M = 1.0\ M\ Ca(NO_3)_2$$

9. This problem gives us the number of liters of solution and the molar concentration of the solution. From the definition of molarity, moles solute/liter solution, we can calculate the number of moles of solute, KI.

$$2.5 \text{ L solution} \times \frac{0.15 \text{ mol KI}}{\text{L solution}} = 0.38 \text{ mol KI}$$

Now use the molar mass of KI to calculate the grams of KI.

$$0.38 \text{ mol KI} \times \frac{166.0 \text{ g KI}}{1 \text{ mol KI}} = 63 \text{ g KI}$$

10.

a. Sulfuric acid produces 2 mol of hydrogen ions and 1 mol of sulfate ions for each mole of sulfuric acid.

$$1 \text{ mol } H_2SO_4(aq) \rightarrow 2 \text{ mol } H^+(aq) + 1 \text{ mol } SO_4^{2-}(aq)$$

A 2.0 M solution of sulfuric acid would contain 2(2.0 mol H^+) per liter, or 4.0 M H^+ total, and 2(1.0 mol SO_4^{2-}) per liter, or 2.0 M SO_4^{2-}.

b. Sodium phosphate produces 3 mol of sodium ions for each mol of sodium phosphate and 1 mol of phosphate ions for each mol of sodium phosphate.

$$1 \text{ mol } Na_3PO_4(aq) \rightarrow 3 \text{ mol } Na^+(aq) + 1 \text{ mol } PO_4^{3-}(aq)$$

A 0.6 M solution of sodium phosphate would contain 3(0.6 mol Na^+) per liter or 1.8 M Na^+ and 1(0.6 mol PO_4^{3-}) per liter, or 0.6 M PO_4^{3-}.

c. Aluminum chloride produces 1 mole of aluminum ions for each mole of aluminum chloride and 3 moles of chloride ions for each mole of aluminum chloride.

$$1 \text{ mol } AlCl_3(aq) \rightarrow 1 \text{ mol } Al^{3+}(aq) + 3 \text{ mol } Cl^-(aq)$$

A 1.5 M solution of aluminum chloride would contain 1(1.5 mol Al^{3+}) per liter, or 1.5 M Al^{3+}, and 3(1.5 mol Cl^-) per liter, or 4.5 M Cl^-.

11. We are given the concentration and the volume of a solution containing KCl and water and are asked for the number of moles of solute, KCl. The number of moles of KCl can be calculated from the definition of molarity, moles solute/liter solution.

$$1.5 \text{ L solution} \times \frac{0.48 \text{ mol KCl}}{\text{L solution}} = 0.72 \text{ mol KCl}$$

12. We are given the concentration of a solution containing NaOH and water, and we are asked for the number of grams of NaOH needed to make 1.50 L of solution. The number of moles of NaOH can be calculated from the definition of molarity, moles solute per liter of solution. The grams of NaOH can be calculated using the molar mass of NaOH.

$$1.50 \text{ L solution} \times \frac{0.650 \text{ mol NaOH}}{\text{L solution}} \times \frac{40.00 \text{ g NaOH}}{\text{mol NaOH}} = 39.0 \text{ g NaOH}$$

13. We are given the concentration of a solution containing K_2SO_4 and water, and we are asked for the number of grams of K_2SO_4 needed to make 250 mL of solution. The number of moles of K_2SO_4 can be calculated from the definition of molarity. The grams of K_2SO_4 can be calculated using the molar mass of K_2SO_4. We will need to convert the given units of volume, milliliters, to liters.

$$250 \text{ mL solution} \times \frac{1 \text{ L solution}}{1000 \text{ mL solution}} \times \frac{0.150 \text{ mol } K_2SO_4}{\text{L solution}} \times \frac{174.27 \text{ g } K_2SO_4}{\text{mol } K_2SO_4} = 6.5 \text{ g } K_2SO_4$$

14. We are given the concentration of a solution containing NaF and water, and we are asked for the number of grams of NaF needed to make 2.0×10^6 L of solution. The number of moles of NaF can be calculated from the definition of molarity. The grams of NaF can be calculated from the molar mass of NaF.

$$2.0 \times 10^6 \text{ L solution} \times \frac{3.0 \times 10^{-6} \text{ mol NaF}}{\text{L solution}} \times \frac{41.99 \text{ g NaF}}{\text{mol NaF}} = 2.5 \times 10^2 \text{ g NaF}$$

15. In this problem we are asked to calculate how much of a concentrated stock solution, which is 12 *M* HCl, is needed to prepare a dilute HCl solution. We will need to know how many moles of HCl are present in 2.5 L of 1.0 *M* HCl, that is, in the dilute solution. Then we need to find a volume of the concentrated solution that contains this same number of moles. We can use this procedure because the number of moles of solute in the dilute solution is the same as the number of moles of solute in the volume of stock solution. Only the volume of water changes. First find the number of moles of HCl that will be present in the dilute solution by multiplying the volume by the molarity.

$$2.5 \text{ L solution} \times \frac{1.0 \text{ mol HCl}}{\text{L solution}} = 2.5 \text{ mol HCl}$$

So the dilute solution will contain 2.5 mol HCl, and the volume of stock solution we need will also contain 2.5 mol HCl. The volume of stock solution multiplied by the molarity of the stock solution equals the number of moles of HCl that will be in the dilute solution.

$$\text{volume of stock soluton} \times \frac{\text{mol HCl in stock solution}}{\text{L stock solution}} = \text{mol HCl in dilute solution}$$

Now substitute values into the equation.

$$V \times \frac{12 \text{ mol HCl}}{\text{L solution}} = 2.5 \text{ mol HCl}$$

Rearrange the equation to isolate V on one side.

$$V = \frac{2.5 \text{ mol HCl}}{12 \text{ mol HCl/L solution}}$$

$$V = 0.21 \text{ L HCl}$$

So to make 2.5 L of 1.0 *M* HCl, use 0.21 L of 12 *M* HCl, and add enough water to bring the total volume to 2.5 L.

A different way to approach this problem is to use the formula

$$M_1 \times V_1 = M_2 \times V_2$$

M_1 represents the molarity of the stock solution; V_1, the volume of the stock solution needed; M_2, the molarity of the dilute solution we wish to make; and V_2, the volume of the dilute solution. In this case we want to know the volume of stock solution needed, so isolate V_1 on one side of the equation by dividing both sides by M_1.

$$M_1 \times V_1 = M_2 \times V_2$$

$$\frac{M_1}{M_1} \times V_1 = V_2 \times \frac{M_2}{M_1}$$

$$V_1 = V_2 \times \frac{M_2}{M_1}$$

Now substitute values into the equation.

$$V_1 = 2.5 \text{ L} \times \frac{1.0\ M}{12\ M}$$

$$V_1 = 0.21 \text{ L}$$

The answer to this problem is the same either way we solve it.

16. We want to know how much concentrated stock solution is needed to make 1855 mL of 0.65 M H_2SO_4. We can use the formula

$$M_1 \times V_1 = M_2 \times V_2$$

M_1 is the molarity of the stock solution, V_1 is the volume of the stock solution, M_2 is the molarity of the dilute solution, and V_2 is the volume of the dilute solution. We want to know how much stock solution is needed, so isolate V_1 on one side of the equation.

$$V_1 = V_2 \times \frac{M_2}{M_1}$$

The volume of the dilute solution, V_2, is given in milliliters. We will need to convert milliliters to liters.

$$1855 \text{ mL} \times \frac{1 \text{ L}}{1000 \text{ mL}} = 1.855 \text{ L}$$

Now substitute values into the equation.

$$V_1 = 1.855 \text{ L} \times \frac{0.65\ M\ H_2SO_4}{18\ M\ H_2SO_4}$$

$$V_1 = 0.067 \text{ L}$$

So 0.067 L of 18 M H_2SO_4 diluted to 1.855 L would produce a 0.65 M H_2SO_4 solution.

17. In this problem a solution of $CaCl_2$ is mixed with a solution of Na_3PO_4. A reaction occurs. We are asked for the number of grams of Na_3PO_4 that will react with the $CaCl_2$. Section 15.6 of your textbook gives five steps for solving problems like this one, so let's follow those same steps here.

Step 1: First, write the balanced molecular equation for this reaction. Because this is a reaction between ionic compounds, we also should write the net ionic equation. The balanced molecular equation is

$$3CaCl_2(aq) + 2Na_3PO_4(aq) \rightarrow Ca_3(PO_4)_2(s) + 6NaCl(aq)$$

The net ionic equation is the solid product, $Ca_3(PO_4)_2$ and the ions that react to form the solid product.

$$3Ca^{2+}(aq) + 2PO_4^{3-}(aq) \rightarrow Ca_3(PO_4)_2(s)$$

Step 2: We need to add just enough PO_4^{3-} to react with all the Ca^{2+}. We need to know how many moles of Ca^{2+} there are in the $CaCl_2$ solution. From the volume and the molarity of the $CaCl_2$ solution, we can calculate the number of moles of Ca^{2+}.

$$V \times M = \text{mol } CaCl_2$$

$$2.5 \text{ L } CaCl_2 \text{ solution} \times \frac{0.10 \text{ mol } CaCl_2}{\text{L } CaCl_2 \text{ solution}} = 0.25 \text{ mol } CaCl_2$$

Each mole of $CaCl_2$ produces 1 mole of Ca^{2+}.

$$0.25 \text{ mol } CaCl_2 \times \frac{1 \text{ mol } Ca^{2+}}{1 \text{ mol } CaCl_2} = 0.25 \text{ mol } Ca^{2+}$$

Step 3: In this problem Ca^{2+} is limiting. We must add just enough PO_4^{3-} to react with all the Ca^{2+}.

Step 4: We need to know how many moles of PO_4^{3-} will react with 0.25 mol Ca^{2+}. We can use the mole ratio from the balanced equation to calculate the moles of PO_4^{3-} that are needed.

$$0.25 \text{ mol } Ca^{2+} \times \frac{2 \text{ mol } PO_4^{3-}}{3 \text{ mol } Ca^{2+}} = 0.17 \text{ mol } PO_4^{3-}$$

So 0.17 mol PO_4^{3-} will react with 0.25 mol Ca^{2+}.

Step 5: We are asked for grams of Na_3PO_4, not moles of PO_4^{3-}, so convert moles of PO_4^{3-} to grams of Na_3PO_4. Each mole of Na_3PO_4 contains 1 mole of PO_4^{3-} ions. We can use the molar mass of Na_3PO_4 to convert from moles Na_3PO_4 to grams Na_3PO_4.

$$0.17 \text{ mol } PO_4^{3-} \times \frac{1 \text{ mol } Na_3PO_4}{1 \text{ mol } PO_4^{3-}} \times \frac{163.94 \text{ g } Na_3PO_4}{1 \text{ mol } Na_3PO_4} = 28 \text{ g } Na_3PO_4$$

18. We want to know how many grams of product, $Fe(OH)_3$, can be produced when two aqueous solutions are mixed together.

Step 1: Write and balance the equation for this reaction.

$$FeCl_3(aq) + 3NaOH(aq) \rightarrow Fe(OH)_3(s) + 3NaCl(aq)$$

From the balanced equation, write the net ionic equation.

$$Fe^{3+}(aq) + 3OH^-(aq) \rightarrow Fe(OH)_3(s)$$

Step 2: We need to know the number of moles of reactant present in each solution. The solution of $FeCl_3$ contains 0.25 mol $FeCl_3$, and each mole of $FeCl_3$ contains 1 mole of Fe^{3+}.

$$0.25 \text{ mol FeCl3} \times \frac{1 \text{ mol } Fe^{3+}}{1 \text{ mol } FeCl_3} = 0.25 \text{ mol } Fe^{3+}$$

We need to know the number of moles of OH^- that are present.

$$V \times M = \text{moles NaOH}$$

$$1.2 \text{ L NaOH solution} \times \frac{0.85 \text{ mol NaOH}}{\text{L NaOH solution}} = 1.0 \text{ mol NaOH}$$

Each mole of NaOH contains 1 mole of OH^-.

$$1.0 \text{ mol NaOH} \times \frac{1 \text{ mol } OH^-}{1 \text{ mol NaOH}} = 1.0 \text{ mol } OH^-$$

Step 3: 0.25 mol Fe^{3+} is mixed with 1.0 mol OH^-. Because each mole of Fe^{3+} requires 3 mol of OH^-, 0.25 mol Fe^{3+} requires 3(0.25 mol OH^-) or 0.75 mol OH^-. Because we have 1.0 mol OH^-, the amount of product that forms is limited by the amount of Fe^{3+}.

Step 4: From the mole ratio, each mole of Fe^{3+} produces 1 mole of $Fe(OH)_3$.

$$0.25 \text{ mol } Fe^{3+} \times \frac{1 \text{ mol } Fe(OH)_3}{1 \text{ mol } Fe^{3+}} = 0.25 \text{ mol } Fe(OH)_3$$

Step 5: We want to know the number of grams of $Fe(OH)_3$, so use the molar mass of $Fe(OH)_3$ to convert from moles to grams.

$$0.25 \text{ mol } Fe(OH)_3 \times \frac{106.87 \text{ g } Fe(OH)_3}{1 \text{ mol } Fe(OH)_3} = 27 \text{ g } Fe(OH)_3$$

19. In this problem we are mixing a solution of HCl of known volume and molarity with a solution of NaOH of known molarity and an unknown volume. We are asked to determine the volume of NaOH that will react with the HCl. We can follow the same steps we have used previously.

Step 1: Write the balanced equation for this reaction.

$$NaOH(aq) + HCl\ (aq) \rightarrow NaCl(aq) + H_2O(l)$$

Now write the net ionic equation.

$$H^+(aq) + OH^-(aq) \rightarrow H_2O(l)$$

Step 2: Calculate the moles of HCl using the formula $V \times M =$ moles.

$$150\text{ mL} \times \frac{1\text{ L HCl}}{1000\text{ mL}} \times \frac{0.25\text{ mol H}^+}{\text{L HCl}} = 0.038\text{ mol H}^+$$

Step 3: This problem requires mixing just enough OH^- to react with all the H^+ that is present. The moles of H^+ determine how much OH^- is to be added. The H^+ ions are limiting.

Step 4: From the net ionic equation we can determine how many moles of OH^- are needed to react with all the H^+.

$$0.038\text{ mol H}^+ \times \frac{1\text{ mol OH}^-}{1\text{ mol H}^+} = 0.038\text{ mol OH}^-$$

Step 5: We now know the moles of OH^- and the molarity. We can use the formula $V \times M$ equals moles to calculate the volume of NaOH. Rearrange the equation to isolate V on one side.

$$V \times \frac{M}{M} = \frac{\text{moles}}{M}$$

$$V = \frac{\text{moles}}{M}$$

Now substitute values into the equation.

$$V = \frac{0.038\text{ mol OH}^-}{0.15\text{ mol NaOH/L NaOH}}$$

$$V = 0.25\text{ L NaOH}$$

So 0.25 L of 0.15 M NaOH will completely react with 150 mL of 0.25 M HCl.

20.

Material	Molar Mass	Equivalents	Equivalent Weight
1.0 mol HCl	36.46	1	36.46
1.0 mol H_2SO_4	98.08	2	49.05
1.0 mol KOH	56.11	1	56.11

21. We want to calculate the normality of a solution of phosphoric acid in water. To do so, we need to know the number of equivalents of phosphoric acid present in 5.6 g phosphoric acid. The equivalent weight of phosphoric acid is

$$\text{equivalent weight } H_3PO_4 = \frac{\text{molar mass } H_3PO_4}{3}$$

$$\text{equivalent weight } H_3PO_4 = \frac{97.99 \text{ g}}{3}$$

$$\text{equivalent weight } H_3PO_4 = 32.66 \text{ g}$$

We can now calculate the equivalents of H_3PO_4 present in 5.6 g H_3PO_4.

$$5.6 \text{ g } H_3PO_4 \times \frac{1 \text{ equiv } H_3PO_4}{32.66 \text{ g } H_3PO_4} = 0.17 \text{ equiv } H_3PO_4$$

The definition of normality is $N = \frac{\text{equiv}}{\text{L}}$. We can use this equation to calculate the normality of the H_3PO_4 solution.

$$N = \frac{0.17 \text{ equiv } H_3PO_4}{125 \text{ mL}} \times \frac{1000 \text{ mL}}{1 \text{ L}} = \frac{1.36 \text{ equiv } H_3PO_4}{\text{L}}$$

This solution is 1.36 N H_3PO_4.

CHAPTER 16

Acids and Bases

INTRODUCTION

In this chapter you will learn about the properties of acids and bases. You know about some of the properties of acids already. Solutions such as lemon juice and vinegar contain acids, and their sour taste is from the acid each contains. You will learn how to determine the properties of acids and how to measure the acidity of a solution.

CHAPTER DISCUSSION

Arrhenius proposed that an acid was anything that produced hydrogen ions in solution, and a base was anything that produced hydroxide ions. Scientists discovered that this definition was too restrictive and that there were other bases besides hydroxide ions. Brønsted and Lowry proposed that an acid was a proton (hydrogen ion) donor, and a base was a proton acceptor. This was a much broader definition of acids and bases.

The general reaction for an acid acting in water is given by the following equation:

$$HA(aq) + H_2O(l) \rightleftarrows H_3O^+(aq) + A^-(aq)$$

The protonated water (H_3O^+) is called the hydronium ion and is the conjugate acid of H_2O, which acts as a base to the acid HA. The remaining part of the acid (A^-) is called the conjugate base.

The reaction of an acid and water is a reversible reaction (see Chapter 17). The proton can be attached to the water molecule or to the conjugate base. The relative attractions of H_2O and A^- for the proton determine whether HA or H_3O^+ predominates. If A^- attracts the proton more strongly than water, the equilibrium lies to the left, and there is relatively little H_3O^+. The acid HA is said to be a weak acid. If the water attracts the proton more strongly than does A^-, the equilibrium lies to the right, and HA is a strong acid because virtually all the protons have dissociated from HA.

The hydrogen ion concentration of a solution is represented by very small numbers. Using scientific notation is a good way to represent small numbers, but calculating and using the pH of a solution is another easy way to represent small numbers. Taking the p of any number means we take the logarithm (log) of that number and multiply the result by −1.

$$pN = (-1) \times \log N$$

When we find the pH of a solution, we take the log of the hydrogen ion concentration in mol/L and multiply by −1.

A log scale can also be used to express hydroxide ion concentration. pOH is calculated the same way as pH, except take the log of the hydroxide ion concentration in mol/L and multiply by −1.

$[H^+]$ and $[OH^-]$ are related to each other through the equilibrium expression for the dissociation of water and the ion product constant K_w (at 25°C).

$$[H^+][OH^-] = K_w = 1.0 \times 10^{-14}$$

If we take the p (−log) of each of the parts of this equation, we have

$$pH + pOH = 14.00$$

This means that, for any aqueous solution at 25°C, the sum of pH and pOH will always equal 14. If we know the pH, we can easily find the pOH and vice versa.

A buffered solution resists large changes in pH even when strong acid or strong base is added. The addition of even small amounts of strong acid or base can greatly lower or raise the pH of a nonbuffered solution. Living systems contain many different kinds of buffers to help keep fluids and tissues at the correct pH, even under stressful conditions.

LEARNING REVIEW

1. Explain the differences between the Arrhenius concept of an acid and a base, and the concept of Brønsted and Lowry.
2. For the reaction of perbromic acid with water:

$$HBrO_4(aq) + H_2O(l) \rightarrow H_3O^+(aq) + BrO_4^-(aq)$$

   a. Which two substances are an acid–conjugate base pair?
   b. Which two substances are a base–conjugate acid pair?
3. Write equations to show what happens when each of the acids below reacts with water.
   a. H_2S
   b. HNO_2
4. Show how acetic acid, $HC_2H_3O_2$, reacts with water by drawing Lewis structures for water and its conjugate acid.
5. When formic acid, HCOOH, is mixed with water, the resulting solution weakly conducts an electric current.
   a. Is formic acid a strong or a weak acid?
   b. Toward which side of the reaction does the equilibrium lie?
   c. Which species is the stronger base, H_2O or $HCOO^-$?
6. Which of the acids below are strong acids and which are weak acids?
   a. HF
   b. H_2SO_4
   c. $HC_2H_3O_2$
   d. $HClO_4$
7. The expression for the dissociation of water is $K_w = [H^+][OH^-]$. Why does liquid water not appear in this expression?
8. All the aqueous solutions below are at a temperature of 25°C.
   a. What is the $[H^+]$ of a solution for which $[OH^-] = 1.5 \times 10^{-6}$ *M*?
   b. What is the $[H^+]$ of a solution for which $[OH^-] = 6.3 \times 10^{-3}$ *M*?
   c. What is the $[OH^-]$ of a solution for which $[H^+] = 3.25 \times 10^{-1}$ *M*?
9. What is the $[H^+]$ of a 0.1 *M* solution of NaOH in water at 25°C?
10. What is the pH of each of the solutions below?
   a. A solution in which $[H^+] = 3.0 \times 10^{-3}$ *M*

b. A solution in which $[H^+] = 5.2 \times 10^{-6}\ M$

c. A solution in which $[OH^-] = 1.4 \times 10^{-1}\ M$

11. What is the pOH of each solution below?

a. $[OH^-] = 4.89 \times 10^{-10}\ M$

b. $[OH^-] = 3.2 \times 10^{-5}\ M$

c. $[H^+] = 1.6 \times 10^{-8}\ M$

12. For any solution, what is the relationship between pH and pOH?

13. The pH of lemon juice is 2.1. What is the pOH?

14. The pOH of black coffee is 9.0. What is the pH?

15. Calculate the $[H^+]$ or $[OH^-]$ for the solutions below.

a. Milk has a pH of 6.9. What is the $[H^+]$?

b. Oven cleaner has a pH of 13.4. What is the $[OH^-]$?

c. A phosphate-containing detergent has a pOH of 4.7. What is the $[OH^-]$?

16. What is the pH of each of the following solutions?

a. 0.02 M HCl

b. $3.5 \times 10^{-3}\ M\ HNO_3$

17. HCl is added to a solution containing H_2CO_3 and $NaHCO_3$.

a. Use an equation to show what would happen to the hydrogen ions from the HCl.

b. Why would the pH of this solution not change drastically when the HCl is added?

ANSWERS TO LEARNING REVIEW

1. Arrhenius's model of acids and bases proposes that acids produce hydrogen ions in aqueous solution while bases produce hydroxide ions. The Brønsted-Lowry model of acids and bases proposes that acids are proton donors while bases are proton acceptors.

2.

a. For the reaction of perbromic acid with water, perbromic acid/perbromate ($HBrO_4/BrO_4^-$) is the acid–conjugate base pair.

b. Water/hydronium ion (H_2O/H_3O^+) is the base–conjugate acid pair.

3. One model of an acid postulates that acids donate protons in aqueous solutions. The proton acceptor is the base, water. In aqueous solutions, acids donate protons to the base water to form the hydronium ion and the conjugate base.

a. $H_2S(aq) + H_2O(l) \rightarrow HS^-(aq) + H_3O^+(aq)$

b. $HNO_2(aq) + H_2O(l) \rightarrow NO_2^-(aq) + H_3O^+(aq)$

4.

$$H_2O + HC_2H_3O_2 \rightarrow [H_3O]^+ + C_2H_3O_2^-$$

5.

a. When formic acid is mixed with water, a weak electric current is generated. An electrical current requires ions. We know that only a few ions are in this solution because the current is weak. So formic acid is a weak acid.

b. Because there are only a few formate and hydronium ions formed, the equilibrium lies toward the left, that is, toward undissociated formic acid.

c. A base can be defined as a proton acceptor. The formate ion has a stronger attraction for hydrogen ions than water does because most of the formic acid is undissociated. The formate ion has pulled the hydrogen ion away from the hydronium ion so $HCOO^-$ is a stronger base than water is.

6. The common strong acids are HCl, HNO_3, H_2SO_4 and $HClO_4$. So, b, H_2SO_4, and d, $HClO_4$, are strong acids. HF and $HC_2H_3O_2$ are weak acids.

7. The concentration of liquid water does not appear in the expression for the dissociation of water because the concentration of water changes very little when dissociation occurs. The concentration of water is considered a constant.

8.

a. The product of $[H^+][OH^-]$ is always equal to 1.0×10^{-14}.

$$K_w = 1.0 \times 10^{-14} = [H^+][OH^-]$$

When we are given either $[H^+]$ or $[OH^-]$, we can calculate the other if we remember that K_w is equal to 1.0×10^{-14}. In this problem, we know that $[OH^-] = 1.5 \times 10^{-6}\ M$, and we want to know $[H^+]$. Rearrange the equation to isolate $[H^+]$ on one side by dividing both sides by $[OH^-]$.

$$K_w = [H^+][OH^-]$$

Divide both sides by $[OH^-]$.

$$\frac{K_w}{[OH^-]} = \frac{[H^+][OH^-]}{[OH^-]}$$

$$[H^+] = \frac{K_w}{[OH^-]}$$

Substitute values into the equation.

$$[H^+] = \frac{1.0 \times 10^{-14}}{1.5 \times 10^{-6}}$$

$$[H^+] = 6.7 \times 10^{-9}\ M$$

b. We are given $[OH^-]$ and asked for $[H^+]$. We can use the K_w expression to find $[H^+]$. Rearrange the equation to isolate $[H^+]$ on one side.

$$K_w = [H^+][OH^-]$$

$$\frac{K_w}{[OH^-]} = \frac{[H^+][OH^-]}{[OH^-]}$$

$$[H^+] = \frac{K_w}{[OH^-]}$$

Substitute values into the equation.

$$[H^+] = \frac{1.0 \times 10^{-14}}{6.3 \times 10^{-3}}$$

$$[H^+] = 1.6 \times 10^{-12}\ M$$

c. We are given $[H^+]$ and asked for $[OH^-]$. Use the K_w expression to find $[OH^-]$. Rearrange the K_w expression to isolate $[OH^-]$ on one side by dividing both sides by $[H^+]$.

$$K_w = [H^+][OH^-]$$

$$\frac{K_w}{[H^+]} = \frac{[H^+][OH^-]}{[H^+]}$$

$$[OH^-] = \frac{K_w}{[H^+]}$$

Now substitute values into the equation.

$$[OH^-] = \frac{1.0 \times 10^{-14}}{3.25 \times 10^{-1}}$$

$$[OH^-] = 3.08 \times 10^{-14}\ M$$

9. When NaOH dissolves in water, each mole of NaOH forms a mole of Na^+ ions and a mole of OH^- ions. A solution that is 0.1 *M* NaOH is also 0.1 *M* OH^-. We can use the K_w expression to find $[H^+]$.

$$K_w = [H^+][OH^-]$$

Rearrange the equation to isolate $[H^+]$ on one side.

$$\frac{K_w}{[OH^-]} = \frac{[H^+][OH^-]}{[OH^-]}$$

$$[H^+] = \frac{K_w}{[OH^-]}$$

Substitute values into the equation, and find $[H^+]$.

$$[H^+] = \frac{1.0 \times 10^{-14}}{0.1}$$

$$[H^+] = 1 \times 10^{-13}\ M$$

The amount of H^+ is very small compared with the amount of OH^-.

10. The pH scale is a way to write very small numbers in a more convenient form. $pH = -\log[H^+]$.

    a. 2.52

    b. 5.28

    c. We are given $[OH^-]$ and are asked for pH. We must find $[H^+]$ before we can find the pH. Use the K_w expression to find $[H^+]$.

$$K_w = [H^+][OH^-]$$

$$[H^+] = \frac{K_w}{[OH^-]}$$

$$[H^+] = \frac{1.0 \times 10^{-14}}{1.4 \times 10^{-1}} = 7.14 \times 10^{-14}$$

    Now that we know $[H^+]$, we can find the pH. The pH of a solution that contains $1.4 \times 10^{-1}\ M\ OH^-$ is 13.15.

11.

    a. The pOH of a solution is equal to $-\log [OH^-]$. The pOH of a solution that contains $4.89 \times 10^{-10}\ M\ OH^-$ is 9.311.

    b. The pOH of a solution is equal to $-\log [OH^-]$. The result is 4.49. The pOH of a solution that contains $3.2 \times 10^{-5}\ M\ OH^-$ is 4.49.

    c. To find the pOH of this solution, we must find the OH^- concentration. Use the K_w expression to find $[OH^-]$.

$$K_w = [H^+][OH^-]$$

$$[OH^-] = \frac{K_w}{[H^+]}$$

$$[OH^-] = \frac{1.0 \times 10^{-14}}{1.6 \times 10^{-8}} = 6.3 \times 10^{-7}\ M$$

    The concentration of OH^- is $6.3 \times 10^{-7}\ M$. The pOH of a solution that contains $1.6 \times 10^{-8}\ M\ H^+$ is 6.20.

12. For all solutions the sum of pH and pOH must equal 14.00. This relationship is derived from the K_w expression, $[H^+][OH^-] = 1.0 \times 10^{-14}$. If we know either the pH or the pOH, we can find the other using this relationship. For example, if the pH of a solution is 4.5, then the pOH is 14.00 minus 4.5, which equals 9.5.

13. If we know either the pH of a solution or the pOH, we can calculate the other. pH and pOH are related by the expression

$$pH + pOH = 14.00$$

    This equation is derived from

$$[H^+][OH^-] = K_w = 1.0 \times 10^{-14}$$

    If the pH of lemon juice is 2.1, then the pOH is

$$pOH = 14.00 - pH$$

$$pOH = 11.9$$

14. pOH and pH are related to each other by the equation

$$pH + pOH = 14$$

If the pOH of black coffee is 9.0, then the pH is

$$pH = 14.00 - pOH$$

$$pH = 5.0$$

15.

a. We want to find $[H^+]$ of milk, and the pH of milk is 6.9. To find the pH of milk from $[H^+]$, we take minus the log of $[H^+]$, so to find $[H^+]$ from pH we must go backward and undo the $\log[H^+]$. We can do this by finding the inverse log of −pH.

$$[H^+] = \text{inverse log}(-pH)$$

$$[H^+] = \text{inverse log}(-6.9) = 1 \times 10^{-7}\ M$$

b. We want to find the $[OH^-]$ of oven cleaner, which has a pH of 13.4. pH and pOH are related to each other by the expression pH + pOH = 14.00.

$$pOH = 14.00 - pH$$

$$pOH = 14.00 - 13.4$$

$$pOH = 0.6$$

$$[OH^-] = \text{inverse log}(-0.6) = 3 \times 10^{-1}\ M$$

c. We want to know the $[OH^-]$ of a phosphate-containing detergent with a pOH of 4.7. We must go backward from pOH to $[OH^-]$.

$$[OH^-] = \text{inverse log}(-4.7) = 2 \times 10^{-5}\ M$$

16.

a. HC1 is a strong acid. This means that, when HC1 is dissolved in water, only H^+ ions and Cl^- ions are present. If a solution is described as 0.02 M HCl, then it actually contains 0.02 M H^+ and 0.02 M Cl^-. Because the pH depends on the hydrogen ion concentration, we can calculate the pH of a solution of HCl if we know the molar concentration of HCl.

$$0.02\ \text{mol/L HCl} = 0.02\ \text{mol/L } H^+ \text{ and } 0.02\ \text{mol/L } Cl^-$$

$$pH = -\log[H^+]$$

$$pH = -\log(0.02)$$

$$pH = 1.7$$

b. 3.5×10^{-3} mol/L HNO_3 dissolved in water produces 3.5×10^{-3} mol/L H^+ and 3.5×10^{-3} mol/L NO_3^-. Use the hydrogen ion concentration to calculate the pH.

$$pH = -\log[H^+]$$

$$pH = -\log(3.5 \times 10^{-3})$$

$$pH = 2.46$$

17.

a. Hydrogen ions produced when HCl is dissolved in water would react with the bicarbonate ion, HCO_3^-, which has a high affinity for hydrogen ions.

$$HCO_3^- + H^+ \rightarrow H_2CO_3$$

b. The pH of a solution is determined by the number of hydrogen ions in a solution. In a solution that contains a mixture of $NaHCO_3$ and H_2CO_3, the $[H^+]$ remains steady. H_2CO_3 is a weak acid, so most of the acid is in the undissociated form, and not much hydrogen ion is produced. But most of the $NaHCO_3$ is dissociated as Na^+ and HCO_3^-. When an outside source of hydrogen ion is added to a buffered solution, the hydrogen ions are removed from the solution by reacting with the conjugate base, HCO_3^-, which has a high affinity for hydrogen ions. The hydrogen ions are removed from the solution, and the pH of the solution does not change much.

CHAPTER 17

Equilibrium

INTRODUCTION

Because atoms and molecules are so tiny, it is hard to imagine what happens when they react and form new products. In this chapter you will learn what is necessary for a reaction to occur, why some reactions stop before all the reactants have been used up, and how to speed up a reaction. Learning how to control chemical reactions has led to many important applications such as new ways to keep food from spoiling.

CHAPTER DISCUSSION

The collision model says that in order for molecules to react with each other, they must first collide. Increases in the temperature and concentration of reactants bring about more collisions, and the rate of reaction increases. The collision model explains many observations about reactions.

Not all molecules that collide react. Colliding molecules must have a minimum amount of energy for a reaction to occur, and this is termed the activation energy. As the temperature increases, the molecules absorb more heat energy, move faster, and when they collide are more likely to meet the activation energy requirement. Therefore, as temperature increases, reaction rate increases. Some substances can cause the reaction rate to increase without increasing the temperature. These substances are called catalysts. Catalysts are useful because they increase the reaction rate without necessitating an increase in temperature or concentration.

Many reactions do not continue until all of the reactants have been converted to products. This is because reactions are reversible. A reversible reaction is one where reactants form products, and products can also form reactants. Both the forward and reverse reactions eventually occur at the same rate, so the concentrations of the products and reactants do not change. Equilibrium is dynamic, though, because both reactions (forward and reverse) are always occurring, but the rates are equal. The reaction system must be in a closed container for this to occur.

The results of measuring the concentrations of reactants and products for many reversible reactions led scientists to formulate the law of chemical equilibrium. This can be stated mathematically as:

$$\mathrm{aA + bB \rightleftarrows cC + dD}$$

$$K = \frac{[\mathrm{C}]^{c}[\mathrm{D}]^{d}}{[\mathrm{A}]^{a}[\mathrm{B}]^{b}}$$

The letter K is called the equilibrium constant, and the amounts of reactants and products are measured in molarity. The equilibrium constant remains the same as long as the temperature is held constant even though the initial concentrations may change. We can write the equilibrium expression for any equation as long as the equation is balanced.

When we write an equilibrium expression for an equilibrium involving solids or pure liquid, we do not include the solid or pure liquid because the concentrations of these are constant.

Thus, the equilibrium expression for

$$2CO_2(g) \rightleftarrows 2CO(g) + O_2(g)$$

is

$$K = \frac{[CO]^2[O_2]}{[CO_2]^2}$$

while the equilibrium expression for

$$Fe_2O_3(s) + 3H_2(g) \rightleftarrows 3H_2O(g) + 2Fe(s)$$

is

$$K = \frac{[H_2O]^3}{[H_2]^3}$$

Le Châtelier's principle helps us predict what happens to systems at equilibrium when conditions are changed. Le Châtelier's principle states that when a system at equilibrium is changed, the system will shift its equilibrium position in order to reduce the change. The most common changes are changes in concentration, volume, and temperature.

When an equilibrium condition is changed by changing the concentration of a reactant or a product, the system will shift to counteract this change. For example, adding a reactant (or removing a product) will cause more product to be formed. Adding a product (or removing a reactant) will cause more reactant to be formed.

Decreasing the volume of a reaction vessel at constant temperature and moles of gas causes an increase in pressure. The gas molecules are in a smaller volume, and they hit the walls more often. Le Châtelier's principle predicts that the system will act to lower the pressure. How can pressure decrease? By having fewer molecules of gas. For example, consider the equation

$$2CO_2(g) \rightleftarrows 2CO(g) + O_2(g)$$

If the volume of a reaction mixture with CO_2, CO, and O_2 is decreased, the system responds by shifting to the left, since three molecules react to form two molecules. K stays the same, but the concentrations change.

Temperature causes the value of K to change, but we can still use Le Châtelier's principle to predict the effect of a change in temperature. To do this, we treat heat just like a product or a reactant. If a reaction is endothermic (absorbs heat), an increase in temperature will shift the reaction to the right. If the reaction is exothermic (gives off heat), an increase in temperature will shift the reaction to the left. Decreasing the temperature has the opposite effect in each case.

Ionic compounds dissolving in water can also reach equilibrium. The equilibrium constant for this type of reaction is called the solubility product and given as K_{sp}. For example, consider the following reaction

$$Al(OH)_3(s) \rightleftarrows Al^{3+}(aq) + 3OH^-(aq)$$

The solubility product is given as

$$K_{sp} = [Al^{3+}][OH^-]^3$$

LEARNING REVIEW

1. Why does an increase in the concentration of reactant cause a reaction to speed up?

2.

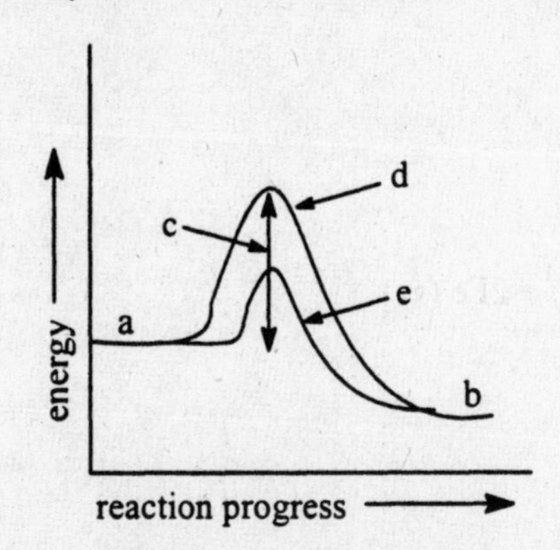

   a. Which letter represents the energy of the products of a reaction?

   b. Which letter represents the energy of the reactants?

   c. Which letter represents the activation energy (E_a) of a reaction?

   d. Which letter represents the catalyzed reaction pathway?

3. What is a catalyst, and how does it work?

4. A beaker of liquid water in a sealed container is allowed to reach equilibrium vapor pressure. What happens to the concentration of water vapor in the beaker from the time the water is placed in the beaker until equilibrium is reached?

5. Which of the following statements about equilibrium are true?

   a. After equilibrium is established, the rate of the forward reaction is greater than the rate of the reverse reaction.

   b. Before equilibrium is reached, the concentration of products increases as time passes.

   c. Before equilibrium is reached, the concentration of reactants increases as time passes.

6. Write equilibrium constant expressions for each of the reactions below.

   a. $I_2(g) + Cl_2(g) \rightleftarrows 2ICl(g)$

   b. $2NO_2(g) + 7H_2(g) \rightleftarrows 2NH_3(g) + 4H_2O(g)$

   c. $4NH_3(g) + 5O_2(g) \rightleftarrows 4NO(g) + 6H_2O(g)$

7. Write the equilibrium expression, and calculate the equilibrium constant, K, for the reaction below at each set of equilibrium concentrations.

$$C_2H_4O_2 \rightleftarrows C_2H_3O_2^- + H^+$$

Experiment	Equilibrium Concentrations			Eq. Expression	K
	$[C_2H_4O_2]$	$[H^+]$	$[C_2H_3O_2^-]$		
I	1.0 M	0.0042 M	0.0042 M		
II	0.50 M	0.0030 M	0.0030 M		
III	2.0 M	0.0060 M	0.0060 M		

8. Why do we not include the concentration of solids or pure liquids in the equilibrium expression?

9. Write the equilibrium constant expression for each of the reactions below.

   a. $Pb(OH)_2(s) \rightleftarrows Pb^{2+}(aq) + 2OH^-(aq)$

   b. $2Sb(s) + 3Cl_2(g) \rightleftarrows 2SbCl_3(g)$

   c. $Fe_2O_3(s) + 3H_2(g) \rightleftarrows 3H_2O(g) + 2Fe(s)$

10. What would happen to the position of the equilibrium when the following changes are made to the equilibrium reaction below?

$$2SO_2(g) + O_2(g) \rightleftarrows 2SO_3(g)$$

   a. SO_2 is removed from the reaction vessel.

   b. SO_3 is added to the reaction vessel.

   c. Oxygen is removed from the reaction vessel.

11. What will happen to the position of the equilibrium when the following changes are made to the reaction below?

$$2HgO(s) \rightleftarrows Hg(l) + O_2(g)$$

   a. Solid $HgO(s)$ is added to the reaction vessel.

   b. The pressure in the reaction vessel is increased by lowering the volume.

12. When the volume of the following mixture of gases is increased, what will be the effect on the equilibrium position?

$$4HCl(g) + O_2(g) \rightleftarrows 2H_2O(g) + 2Cl_2(g)$$

13. Predict the effect of decreasing the volume of the container on the position of each equilibrium below.

   a. $SiF_4(g) + 2H_2O(g) \rightleftarrows SiO_2(s) + 4HF(g)$

   b. $2H_2(g) + 2NO(g) \rightleftarrows 2H_2O(g) + N_2(g)$

   c. $C(s) + H_2O(g) \rightleftarrows CO(g) + H_2(g)$

14. Predict the effect of increasing the temperature on the position of each equilibrium below.

   a. $H_2(g) + Cl_2(g) \rightleftarrows 2HCl(g) + \text{heat}$ exothermic
   b. $2NH_3(g) + \text{heat} \rightleftarrows N_2(g) + 3H_2(g)$ endothermic
   c. $CO_2(g) + H_2(g) + \text{heat} \rightleftarrows CO(g) + H_2O(g)$ endothermic

15. If the equilibrium constant for the reaction below is 51.47, the concentration of HI is 0.50 M, and the concentration of H_2 is 0.069 M, what is the concentration of I_2?

$$H_2(g) + I_2(g) \rightleftarrows 2HI(g)$$

16. Write the balanced equation describing the dissolution of the solids below. Then write the K_{sp} expression.

   a. $Ca_3(PO_4)_2$

   b. FeS

c. $Al(OH)_3$

17. The solubility of $PbSO_4(s)$ is 1.3×10^{-4} mol/L at 25°C. What is the K_{sp} of $PbSO_4(s)$?

18. Copper(II) sulfide has a K_{sp} of 8.0×10^{-45}. What is the solubility of $CuS(s)$ in water at 25°C?

ANSWERS TO LEARNING REVIEW

1. The collision model says for reactions to occur the reactants must first collide with each other. Higher concentrations of reactants cause the reaction rate to speed up because the number of collisions increases as the concentration of reactants increases.

2..

   a. Letter b represents the energy of the products. In this part of the graph, the reactants have achieved the activation energy, and products have formed.

   b. Letter a represents the energy of the reactants. The average energy of the reactants is lower than the activation energy.

   c. Letter c represents the activation energy, the minimum amount of energy needed for a reaction to occur.

   d. Letter e represents the catalyzed reaction pathway. A catalyst lowers the activation energy for a reaction.

3. A catalyst is a substance added to the reaction that causes the reaction to speed up. Catalysts are not used up during the reaction. They are not reactants and do not form products. A catalyst works by providing a new path for the reaction, which lowers the activation energy for that reaction.

4. When water is first sealed in the beaker, the level of water in the beaker decreases as water from the beaker enters the vapor phase. After some period of time, the level of water stops decreasing and stays at the same level. At this point a balance occurs between the processes of evaporation and condensation. The system is at equilibrium, and the concentration of water vapor does not change.

5.

   a. This statement is false. At equilibrium the rate of forward reaction equals the rate of the reverse reaction.

   b. This statement is true. When reactants are mixed, they continue to react to form product. The amount of product increases until equilibrium is reached.

   c. This statement is false. The concentration of reactants decreases until equilibrium is reached, at which point the concentration of reactants remains constant.

6. When writing equilibrium expressions, coefficients become powers. Products appear in the numerator and reactants in the denominator.

   a. $K = \dfrac{[ICl]^2}{[I_2][Cl_2]}$

   b. $K = \dfrac{[H_2O]^4[NH_3]^2}{[NO_2]^2[H_2]^7}$

   c. $K = \dfrac{[NO]^4[H_2O]^6}{[NH_3]^4[O_2]^5}$

7.

Experiment	Equilibrium Concentrations			Eq. Expression	K
	$[C_2H_4O_2]$	$[H^+]$	$[C_2H_3O_2^-]$		
I	1.0M	0.0042M	0.0042M	$K = \frac{[H^+][C_2H_3O_2^-]}{[C_2H_4O_2]}$	1.8×10^{-5}
II	0.50M	0.0030M	0.0030M	$K = \frac{[H^+][C_2H_3O_2^-]}{[C_2H_4O_2]}$	1.8×10^{-5}
III	2.0M	0.0060M	0.0060M	$K = \frac{[H^+][C_2H_3O_2^-]}{[C_2H_4O_2]}$	1.8×10^{-5}

8. The concentrations of pure solids and liquids are constant and do not change. They are not shown in the equilibrium expression.

9.

a. $K_{sp} = [Pb^{2+}][OH^-]^2$

b. $K = \frac{[SbCl_3]^2}{[Cl_2]^3}$

c. $K = \frac{[H_2O]^3}{[H_2]^3}$

10.

a. The equilibrium will shift to the left and begin producing SO_2.

b. The equilibrium will shift to the left and begin consuming SO_3.

c. The equilibrium will shift to the left and begin producing O_2.

11.

a. Pure solids and liquids have no effect on the equilibrium, so adding or removing HgO(*s*) or Hg(*l*) will not affect the position of the equilibrium.

b. The equilibrium will shift toward the left to reduce the pressure inside the system by consuming oxygen gas.

12. On the left side there are five gaseous molecules, and on the right there are four. When the volume is increased, the pressure decreases, and the equilibrium will shift toward the side that increases the pressure, to the left.

13. Decreasing the volume causes an increase in pressure. The equilibrium shifts in the direction to relieve the pressure, or toward the side with fewer numbers of gaseous molecules.

a. The equilibrium would shift toward the left because the left has three gaseous molecules while the right has four gaseous molecules.

b. The equilibrium would not shift in either direction because the number of gaseous molecules on the right and the left is the same.

c. The equilibrium would shift toward the left because the left has one gaseous molecule and the right has two gaseous molecules. Solid carbon on the left does not affect the equilibrium.

14. For exothermic reactions, treat heat energy as a product of the reaction. If heat energy is added to the system by raising the temperature, then the equilibrium shifts to the left to consume the heat energy. For endothermic reactions, treat heat energy as a reactant. If heat energy is added to the system by raising the temperature, then the equilibrium shifts to the right to consume the heat energy.

    a. This is an exothermic reaction. Raising the temperature would shift the equilibrium to the left to consume the energy.

    b. This is an endothermic reaction. Raising the temperature would shift the equilibrium to the right to consume excess energy.

    c. This is an endothermic reaction. Raising the temperature would shift the equilibrium to the right to consume the energy.

15. First write the equilibrium expression for this reaction.

$$K = \frac{[\mathrm{HI}]^2}{[\mathrm{H_2}][\mathrm{I_2}]}$$

Rearrange this equation to isolate I_2 on one side. Divide both sides by $[HI]^2$.

$$\frac{K}{[\mathrm{HI}]^2} = \frac{[\mathrm{HI}]^2}{[\mathrm{HI}]^2} \times \frac{1}{[\mathrm{H_2}][\mathrm{I_2}]}$$

$$\frac{K}{[\mathrm{HI}]^2} = \frac{1}{[\mathrm{H_2}][\mathrm{I_2}]}$$

Now, multiply both sides by $[H_2]$.

$$\frac{K[\mathrm{H_2}]}{[\mathrm{HI}]^2} = \frac{1}{[\mathrm{I_2}]}$$

Take the inverse of both sides.

$$[\mathrm{I_2}] = \frac{[\mathrm{HI}]^2}{K[\mathrm{H_2}]}$$

Substitute values into the equation and find $[I_2]$.

$$[\mathrm{I_2}] = \frac{[0.25\ \mathrm{mol/L}]^2}{51.47[0.069\ \mathrm{mol/L}]}$$

$$[\mathrm{I_2}] = 0.070\ \mathrm{mol/L}$$

16. The K_{sp} expression does not include the concentration of the solid salt, only the ions in solution.

    a. $K_{sp} = [Ca^{2+}]^3[PO_4^{3-}]^2$

    b. $K_{sp} = [Fe^{2+}][S^{2-}]$

    c. $K_{sp} = [Al^{3+}][OH^-]^3$

17. This problem asks us to calculate the K_{sp} for $PbSO_4$ given the solubility of $PbSO_4$. When $PbSO_4$ dissolves in water, lead ions and sulfate ions are released into the aqueous environment.

$$\mathrm{PbSO_4}(s) \rightleftarrows \mathrm{Pb^{2+}}(aq) + \mathrm{SO_4^{2-}}(aq)$$

The K_{sp} expression for the reaction is

$$K_{sp} = [Pb^{2+}][SO_4^{2-}]$$

We need to know the molar concentrations of Pb^{2+} and SO_4^{2-} to find the K_{sp}. We know that the solubility of $PbSO_4$ is 1.3×10^{-4} mol/L. This means that 1.3×10^{-4} mol $PbSO_4$ dissolves per liter of solution. Each 1.3×10^{-4} mol $PbSO_4$ produces 1.3×10^{-4} mol Pb^{2+} and 1.3×10^{-4} mol SO_4^{2-}. The concentration of Pb^{2+} is 1.3×10^{-4} mol/L, and the concentration of SO_4^{2-} is 1.3×10^{-4} mol/L. We can use these concentrations to calculate the K_{sp}.

$$K_{sp} = [Pb^{2+}][SO_4^{2-}]$$

$$K_{sp} = (1.3 \times 10^{-4}\ \text{mol/L})(1.3 \times 10^{-4}\ \text{mol/L})$$

$$K_{sp} = 1.7 \times 10^{-8}\ \text{mol}^2/\text{L}^2$$

The units for K_{sp} are usually omitted, so the K_{sp} would be reported as 1.7×10^{-8}.

18. This problem asks us to find the solubility of copper(II) sulfide. We are given the K_{sp}, which is 8.0×10^{-45}. When CuS dissolves in water, each mole of CuS that dissolves produces one mole of Cu^{2+} and one mole of S^{2-}.

$$CuS(s) \rightleftarrows Cu^{2+}(aq) + S^{2-}(aq)$$

The K_{sp} expression for the dissolution of CuS in water is

$$K_{sp} = [Cu^{2+}][S^{2-}]$$

$x \dfrac{\text{mol}}{\text{L}}$ of CuS(s) dissociates into $x \dfrac{\text{mol}}{\text{L}}$ $Cu^{2+}(aq)$ and $x \dfrac{\text{mol}}{\text{L}}$ $S^{2-}(aq)$.

At equilibrium $[Cu^{2+}] = x \dfrac{\text{mol}}{\text{L}}$ and $[S^{2-}] = x \dfrac{\text{mol}}{\text{L}}$.

We can substitute these values into the equilibrium expression.

$$K_{sp} = 8.0 \times 10^{-45} = [Cu^{2+}][S^{2-}] = (x)(x) = x^2$$

We can say that

$$x^2 = 8.0 \times 10^{-45}$$

$$x = \sqrt{8.0 \times 10^{-45}} = 8.9 \times 10^{-23}\ \text{mol/L}$$

The solubility of CuS is 8.9×10^{-23} mol/L.

CHAPTER 18

Oxidation-Reduction Reactions and Electrochemistry

INTRODUCTION

There are many important oxidation–reduction reactions. These reactions are characterized by electron transfer. One of the most annoying and costly of the oxidation–reduction reactions is the rusting of automobile bodies. In this chapter you will learn what actually happens during an oxidation–reduction reaction and what means are available to keep your car from rusting.

CHAPTER DISCUSSION

Reactions between metals and nonmetals involve the transfer of electrons. When the metal Li reacts with Br_2, the result is the ionic compound LiBr.

$$2Li + Br_2 \rightarrow 2LiBr$$

Both Li and Br_2 began as neutral species, but after the reaction both were ions. Li became the Li^+ cation, and Br_2 became the Br^- anion. Electrons must have been transferred between lithium and bromine for the reaction to have occurred. Reactions of this type are called oxidation–reduction reactions or redox reactions. In this reaction, Li has lost an electron to become the Li^+ cation; this process is called oxidation. Each Br atom has gained an electron to form Br^-; this process is called reduction. In every oxidation–reduction reaction one species is oxidized, and another is reduced.

In some redox reactions it is not easy to tell which species has been oxidized and which has been reduced. The assignment of oxidation states or oxidation numbers can help you determine which species is oxidized and which reduced and whether a reaction is really a redox reaction or not. An oxidation number is an imaginary number assigned to each element in a chemical reaction. The number comes from the charge an element would have if it were an ion. How the electrons are assigned to the atoms is governed by a set of rules, but is basically determined by the electronegativity of the atom. Some species are easy to assign oxidation states to. All the metals that form ions with a 1+ charge have oxidation states of +1 (like K^+, Na^+, etc.). Many atoms in chemical reactions are not ions but are covalently bonded to other atoms; that is, the electrons are shared between two atoms. Assign the electrons as though the atom were ions. The most electronegative atom is assigned both the shared electrons. Because the molecules are electrically neutral, the sum of the oxidation states of all the atoms must be zero.

In Chapter 6 you learned how to balance simple chemical reactions by inspection. Balancing redox reactions is more difficult and can rarely be done by inspection. Another method is needed for balancing these reactions. One method for balancing redox reactions is called the half-reaction method. A half-reaction is part of a complete chemical equation. In a redox reaction, there is always a reduction half-reaction and an oxidation half-reaction. We can write each of them separately and balance the difference in oxidation states on each side of the equation by adding electrons to either the right or the left side. Section 18.4 in your textbook gives some general steps to be used when balancing redox reactions and five specific steps to use when balancing redox reactions which take place in acidic solution. Questions 7 and 8 in Learning Review will test your ability to solve these types of problems.

Chemical and electrical energy can be interchanged. The study of this interchange is called electrochemistry. During a redox reaction, electrons are transferred whenever the reactants collide in solution. It is not possible to use the electron transfer to generate electrical energy under these circumstances. In order to harness the energy of a redox reaction, it is necessary to physically separate the two half-reactions in two separate containers that are connected by a wire. The electrons that are transferred between the oxidation half-reaction and the reduction half-reaction travel along the wire and produce an electrical current.

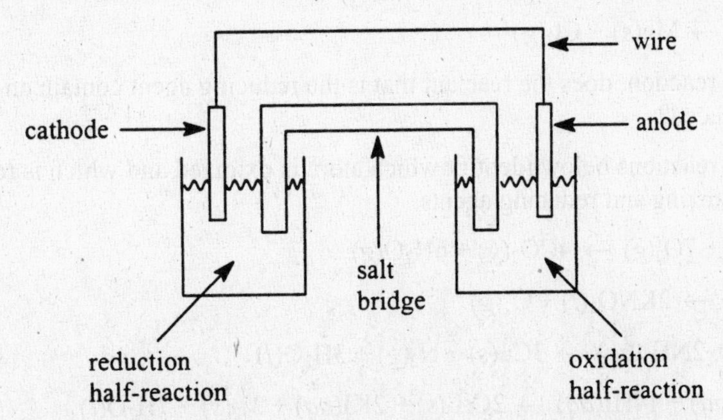

A reaction between two half-cells connected by only a wire will not happen unless there is another connection between the two containers that allows ions to flow freely back and forth. As electrons leave one container and travel along the wire to the other container, differences in charges in the two containers will occur. The container with the oxidation half-reaction will build up a negative charge from the gain of electrons. The extra connection that contains ions allows negative ions to travel to the container losing electrons and positive ions to travel to the container gaining electrons, so the net charge in each container is zero. This connection is called a salt bridge. The current that is produced in a cell such as this can be used to do useful work and is the principle upon which batteries are made. Cells powered by two separate half-reactions connected by a wire and by some type of connection to allow ion exchange are called galvanic cells. The electrode where electrons are lost is called the anode, and the electrode where electrons are gained is called the cathode. A battery is a galvanic cell.

LEARNING REVIEW

1. For each of the partial reactions, decide whether oxidation or reduction is occurring.

   a. $Li \rightarrow Li^{+} + e^{-}$

   b. $Br_2 + 2e^{-} \rightarrow 2Br^{-}$

   c. $S^{2-} \rightarrow S + 2e^{-}$

2. For each reaction below, identify which element is oxidized and which is reduced.

   a. $Ca(s) + I_2(g) \rightarrow CaI_2(s)$

   b. $2K(s) + S(s) \rightarrow K_2S(s)$

   c. $6Na(s) + N_2(g) \rightarrow 2Na_3N(s)$

3. Determine the oxidation states for each element in the substances below.

   a. CH_4

   b. SO_4^{2-}

   c. $NaHCO_3$

d. N_2O_5

e. HIO_4

4. Determine the oxidation state for each element in the reactions below.

a. $4Fe(s) + 3O_2(g) + 12HCl(aq) \rightarrow 4FeCl_3(aq) + 6H_2O(l)$

b. $Zn(s) + 2AgNO_3(aq) \rightarrow Zn(NO_3)_2(aq) + 2Ag(s)$

c. $MgCl_2(l) \rightarrow Mg(s) + Cl_2(g)$

5. During a redox reaction, does the reactant that is the reducing agent contain an element that is oxidized or reduced?

6. For each of the reactions below identify which atom is oxidized and which is reduced, and identify the oxidizing and reducing agents.

a. $2C_2H_6(g) + 7O_2(g) \rightarrow 4CO_2(g) + 6H_2O(g)$

b. $2KNO_3(l) \rightarrow 2KNO_2(l) + O_2(g)$

c. $3CuO(s) + 2NH_3(aq) \rightarrow 3Cu(s) + N_2(g) + 3H_2O(l)$

d. $K_2Cr_2O_7(aq) + 14HI(aq) \rightarrow 2CrI_3(s) + 2KI(aq) + 3I_2(s) + 7H_2O(l)$

7. Balance each of the reactions below by the half-reaction method.

a. $Zn(s) + Cu^{2+}(aq) \rightarrow Zn^{2+}(aq) + Cu(s)$

b. $Re^{5+}(aq) + Sb^{3+}(aq) \rightarrow Re^{4+}(aq) + SbS^{+}(aq)$

8. Each of the reactions below occurs in acidic solution. Balance each one by the half-reaction method.

a. $H_2S(aq) + NO_3^-(aq) \rightarrow S_8(s) + NO(g)$

b. $H_5IO_6(aq) + I^-(aq) \rightarrow I_2(s)$

c. $Cr_2O_7^{2-}(aq) + Sn^{2+}(aq) \rightarrow Sn^{4+}(aq) + Cr^{3+}(aq)$

d. $I_2(s) + NO_3^-(aq) \rightarrow IO_3^-(aq) + NO_2(g)$

9. Normally, when a redox reaction occurs, no useful work is produced. How can a redox reaction be made to perform useful work?

10. Label what is needed to complete the electrical circuit and allow the redox reaction to proceed.

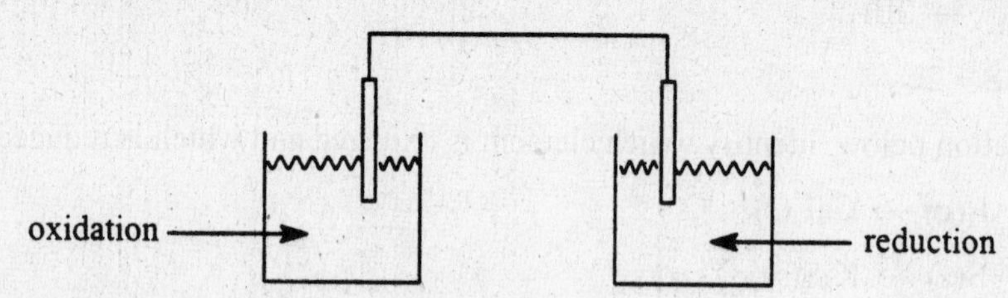

11. Briefly explain how the lead storage battery works.

12. Aluminum metal easily loses electrons to form Al_2O_3. How can aluminum metal be produced from its oxide?

ANSWERS TO LEARNING REVIEW

1.

a. Lithium metal loses an electron. This is oxidation.

b. Bromine gains an electron. This is reduction.

c. The sulfide ion loses two electrons. This is oxidation.

2.

a. Calcium is a metal from Group 2 and forms Ca^{2+} cations. Calcium metal loses two electrons. This is oxidation. Halogens such as I_2 form anions. Each atom in a molecule of 12 gains one electron to form $2I^-$. This is reduction.

b. Potassium metal from Group 1 forms K^+ cations. Potassium loses one electron, so this is oxidation. Sulfur from Group 6 forms the S^{2-} anion. Sulfur gains two electrons, so it is reduced.

c. Sodium metal forms Na^+ cations. Sodium loses electrons, so it is oxidized. Nitrogen from Group 5 forms the N^{3-} anion. Each nitrogen atom gains three electrons, so it is reduced.

3. If you have trouble assigning oxidation states, look at the rules that are found in Section 18.2 of your textbook.

a. Rule 4 says that hydrogen bonded to a nonmetal such as carbon will have an oxidation state of 1+. There are four hydrogen atoms, so carbon must be 4− so that the sum of the charges is zero.

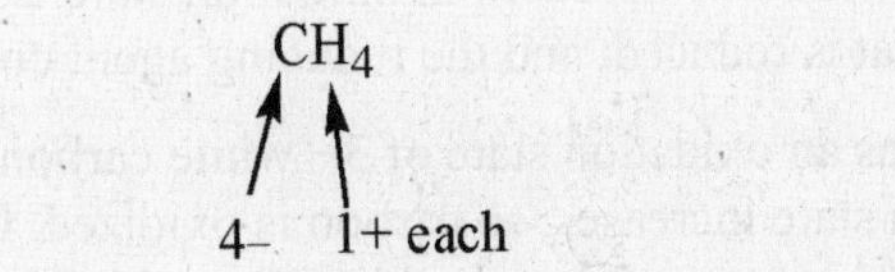

b. Rule 3 says that oxygen is usually 2−. There are four oxygen atoms, so sulfur must be 6+. The sum of the charges must be 2− because the charge on the sulfate ion is 2−.

SO_4^{2-}

6+ 2− each

c. Rule 2 says that the charge on Group 1 ions is 1+, so Na^+ is 1+. Rule 4 says that hydrogen is 1+ when covalently bonded to nonmetals. In this molecule, the hydrogen atom is covalently bonded to the CO_3 part of the molecule, so the oxidation state of hydrogen is 1+. Rule 3 says that oxygen is usually 2−. There are three of them. Carbon must be 6 − (1 + 1), which is 4+.

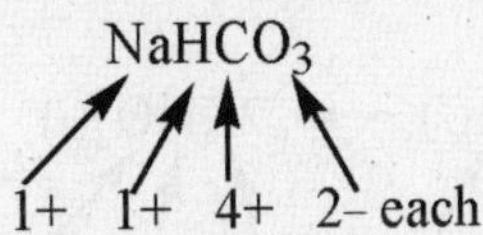

d. Rule 3 says that oxygen is usually 2−. There are five oxygen atoms. There are two nitrogen atoms, so each nitrogen atom must be 5+ to counterbalance the five oxygen atoms.

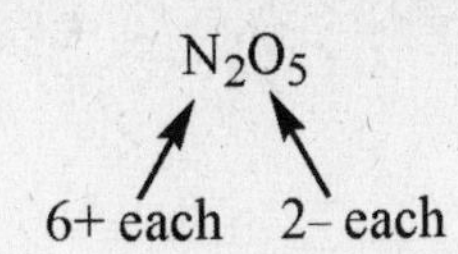

e. Rule 4 says that hydrogen is usually 1+. Rule 3 says that oxygen is usually 2−. If hydrogen is plus one and the four oxygen atoms are 2− each, then iodine must be 7+, so the sum is zero.

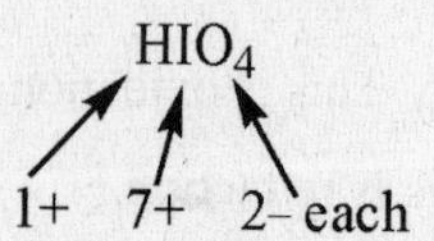

4.

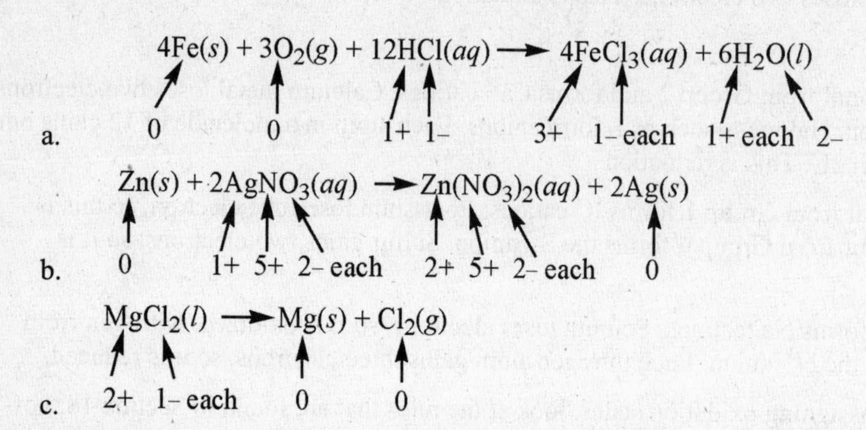

5. The reducing agent contains an element that is oxidized, so the element that loses electrons during oxidation furnishes the electrons needed for reduction.

6. Use changes in oxidation state to determine which element is oxidized and which is reduced. Remember that the element that *increases* in oxidation state is oxidized. The oxidizing agent contains the element that is reduced, and the reducing agent contains the element that is oxidized.

a. Carbon in C_2H_6 has an oxidation state of 3− while carbon in CO_2 has an oxidation state of 4+. The oxidation state increases, so carbon is oxidized. C_2H_6 is the reducing agent. Oxygen in O_2 has an oxidation state of zero, and in CO_2 and in H_2O oxygen has an oxidation state of 2−. The oxidation state of oxygen decreases, so oxygen is reduced. Oxygen gas is the oxidizing agent.

$$2C_2H_6(g) + 7O_2(g) \rightarrow 4CO_2(g) + 6H_2O(g)$$

3− each 1+ each 0 4+ 2− each 1+ each 2−

b. Nitrogen in KNO_3 has an oxidation state of 5+ while nitrogen in KNO_2 has an oxidation state of 3+. The oxidation state decreases from 5+ to 3+, so nitrogen is reduced. KNO_3 is the oxidizing agent. Oxygen in KNO_3 has an oxidation state of 2− while molecular oxygen has an oxidation state of zero. The oxidation state of oxygen increases, so oxygen is oxidized. KNO_3 is the reducing agent.

$$2KNO_3(l) \rightarrow 2KNO_2(l) + O_2(g)$$

1+ 5+ 2− each 1+ 3+ 2− each 0 each

c. Copper in CuO has an oxidation state of 2+ while copper metal has an oxidation state of zero. The oxidation state decreases, so copper is reduced. CuO is the oxidizing agent. Nitrogen in NH_3 has an oxidation state of 3−, while molecular nitrogen has an oxidation state of zero. The oxidation state increases, so nitrogen is oxidized. NH_3 is the reducing agent.

$$3CuO(s) + 2NH_3(aq) \longrightarrow 3Cu(s) + N_2(g) + 3H_2O(l)$$

2+ 2– 3– 1+ each 0 0 each 1+ each 2–

d. Chromium in $K_2Cr_2O_7$ has an oxidation state of 6+ while chromium in CrI_3 has an oxidation state of 3+. The oxidation state of chromium decreases, so chromium is reduced. $K_2Cr_2O_7$ is the oxidizing agent. Iodine in HI has an oxidation state of 1− while iodine molecules have an oxidation state of zero. Note that some of the iodine does not change oxidation state. The iodine atoms in CrI_3 and KI both have oxidation states of 1−. Because the oxidation state of some of the iodine has increased, iodine is said to be oxidized. HI is the reducing agent.

$$K_2Cr_2O_7(aq) + 14HI(aq) \longrightarrow 2CrI_3(s) + 3I_2(s) + 7H_2O(l) + 2KI(aq)$$

1+ each 6+ each 2– each 1+ 1– 3+ 1– each 0 each 1+ each 2– 1+ 1–

7.

a. To balance redox reactions that do not occur in acid solution, follow the general steps given in Section 18.4 of your textbook.

Write individual oxidation and reduction half-reactions.

$$Zn(s) + Cu^{2+}(aq) \rightarrow Zn^{2+}(aq) + Cu(s)$$

The oxidation state of zinc increases from 0 to 2+. Zinc is oxidized.

$$Zn \rightarrow Zn^{2+} \quad \text{oxidation half-reaction}$$

The oxidation state of copper decreases from 2+ to 0, so copper is reduced.

$$Cu^{2+} \rightarrow Cu \quad \text{reduction half-reaction}$$

The Cu^{2+} ion gains electrons to produce copper metal. Two electrons are gained by the Cu^{2+} ion. To balance the 2+ charge on the left side of the reduction half-reaction, add two electrons to the left side.

$$2e^- + Cu^{2+} \rightarrow Cu$$

Zinc metal loses two electrons to become the Zn^{2+} ion. To balance the 2+ charge on the right side of the oxidation half-reaction, add two electrons to the right side.

$$Zn \rightarrow Zn^{2+} + 2e^-$$

Balance the number of atoms in each half-reaction. In this reaction the number of copper atoms is the same on both sides, so the coefficients of solid copper and Cu^{2+} ion do not need to be adjusted.

Zinc metal loses two electrons to become the Zn^{2+} ion. The number of zinc atoms is the same on both sides, so the coefficients of solid zinc and Zn^{2+} ion do not need to be adjusted.

In a balanced oxidation–reduction reaction the number of electrons lost must equal the number of electrons gained. In the oxidation half-reaction two electrons are lost, and in the reduction half-reaction two electrons are gained. Because the number of electrons is the same in both half-reactions, no adjustment is needed to the number of electrons.

Add the two half-reactions together.

$$2e^- + Cu^{2+} \rightarrow Cu$$
$$Zn \rightarrow Zn^{2+} + 2e^-$$
$$2e^- + Cu^{2+} + Zn \rightarrow Cu + Zn^{2+} + 2e^-$$

Now cancel the electrons that appear on both sides to give the overall reaction.

$$Cu^{2+}(aq) + Zn(s) \rightarrow Cu(s) + Zn^{2+}(aq)$$

We can check our work to make sure the elements and charges are the same on both sides. There is one copper and one zinc on each side, and the charge is 2+ on each side, so the equation is balanced.

	$Cu^{2+}(aq)$	+ $Zn(s)$	$\rightarrow$	$Cu(s)$	+ $Zn^{2+}(aq)$
elements	1 Cu	1 Zn	$\rightarrow$	1 Cu	1 Zn
charge	2+		$\rightarrow$	2+	

b. Write the individual oxidation and reduction half-reactions.

$$Re^{5+}(aq) + Sb^{3+}(aq) \rightarrow Re^{4+}(aq) + Sb^{5+}(aq)$$

The oxidation state of rhenium decreases from 5+ to 4+. Rhenium is reduced.

$$Re^{5+} \rightarrow Re^{4+} \quad \text{reduction half-reaction}$$

The oxidation state of antimony increases from 3+ to 5+. Antimony is oxidized.

$$Sb^{3+} \rightarrow Sb^{5+} \quad \text{oxidation half-reaction}$$

The Re^{5+} ion gains an electron to become the Re^{4+} ion. To balance the 1+ charge on the left side of the reduction half-reaction, add one electron to the left side.

$$e^- + Re^{5+} \rightarrow Re^{4+}$$

The Sb^{3+} ion loses two electrons to become the Sb^{5+} ion. To balance the 2+ charge on the right side of the oxidation half-reaction, add two electrons to the right side.

$$Sb^{3+}(aq) \rightarrow Sb^{5+}(aq) + 2e^-$$

The number of rhenium atoms and antimony atoms is the same on both sides, so the coefficients of Re^{5+}, Re^{4+}, Sb^{3+}, and Sb^{5+} do not need to be adjusted.

In a balanced redox reaction the number of electrons gained and lost must be equal, so multiply the reduction half-reaction by two so that both half-reactions transfer two electrons.

$$2(e^- + Re^{5+} \rightarrow Re^{4+})$$
$$2e^- + 2Re^{5+} \rightarrow 2Re^{4+}$$

Now add the two half-reactions together.

$$2e^- + 2Re^{5+} \rightarrow 2Re^{4+}$$
$$Sb^{3+} \rightarrow Sb^{5+} + 2e^-$$
$$2e^- + 2Re^{5+} + Sb^{3+} \rightarrow 2Re^{4+} + Sb^{5+} + 2e^-$$

Cancel the electrons that appear on both sides of the equation to give the overall balanced reaction.

$$2Re^{5+}(aq) + Sb^{3+}(aq) \rightarrow 2Re^{4+}(aq) + Sb^{5+}(aq)$$

We can check our work to make sure the elements and charges are the same on both sides. There are two rheniums and one antimony on both sides, and the charge is 13+ on each side, so the equation is balanced.

	$2Re^{5+}(aq) + Sb^{3+}(aq)$		$\rightarrow$	$2Re^{4+}(aq) + Sb^{5+}(aq)$	
elements	2 Re	1 Sb	$\rightarrow$	2 Re	1 Sb
charge	13+		$\rightarrow$	13+	

8. When balancing oxidation–reduction reactions in acidic solution, use the five steps given in Section 18.4 of your textbook.

a. Step 1: Write equations for the oxidation and reduction half-reactions.

$$H_2S(aq) + NO_3^-(aq) \longrightarrow S_8(s) + NO(g)$$

Oxidation states: H_2S: H 1+ each, S 2−; NO_3^-: N 5+, O 2− each; S_8: 0 each; NO: N 2+, O 2−

Sulfur in H_2S loses two electrons to become elemental sulfur, so sulfur is oxidized. The oxidation half-reaction is

$$H_2S \rightarrow S_8 \quad \text{oxidation half-reaction}$$

Nitrogen in NO_3^- gains three electrons to become NO, so nitrogen is reduced. The reduction half-reaction is

$$NO_3^- \rightarrow NO \quad \text{reduction half-reaction}$$

Step 2a: Balance all elements except hydrogen and oxygen.

The right side of the oxidation half-reaction has eight sulfur atoms, so we will need eight on the left.

$$8H_2S \rightarrow S_8$$

The reduction half-reaction contains one nitrogen atom on both sides, so no adjustment is needed.

Step 2b: Balance the oxygen atoms using H_2O.

The oxidation half-reaction contains no oxygen atoms. The reduction half-reaction has three oxygen atoms on the left and only one on the right, so add two molecules of H_2O to the right side.

$$NO_3^- \rightarrow NO + 2H_2O$$

Step 2c: Balance hydrogens using H^+.

The oxidation half-reaction has sixteen hydrogens on the left and none on the right, so add $16H^+$ to the right side.

$$8H_2S \rightarrow S_8 + 16H^+$$

The reduction half-reaction has four hydrogens on the right and none on the left, so add $4H^+$ to the left side.

$$4H^+ + NO_3^- \rightarrow NO + 2H_2O$$

Step 2d: Balance the charge using electrons.

To balance the sixteen positive charges on the right side of the reduction half-reaction, add sixteen electrons to the right side.

$$8H_2S \rightarrow S_8 + 16H^+$$

charge 0 → 16+

$$8H_2S \rightarrow S_8 + 16H^+ + 16e^-$$

To balance the three positive charges on the left side of the oxidation half-reaction, add three electrons to the left side.

$$4H^+ + NO_3^- \rightarrow NO + 2H_2O$$

charge 3+ → 0

$$3e^- + 4H^+ + NO_3^- \rightarrow NO + 2H_2O$$

Step 3: The oxidation half-reaction transfers sixteen electrons, and the reduction half-reaction transfers three electrons, so we must equalize the number of electrons transferred. Multiply the oxidation half-reaction by three and the reduction half-reaction by sixteen.

$$16(3e^- + 4H^+ + NO_3^- \rightarrow NO + 2H_2O)$$

$$48e^- + 64H^+ + 16NO_3^- \rightarrow 16NO + 32H_2O$$

$$3(8H_2S \rightarrow S_8 + 16H^+ + 16e^-)$$

$$24H_2S \rightarrow 3S_8 + 48H^+ + 48e^-$$

Step 4: Now, add the half-reactions together, and cancel species that appear on both sides. The 48 electrons appear on both sides and cancel. All 48 hydrogen ions cancel on the right, and 48 cancel on the left, leaving 16 hydrogen ions on the left.

$$24H_2S \rightarrow 3S_8 + 48H^+ + 48e^-$$

$$48e^- + 64H^+ + 16NO_3^- \rightarrow 16NO + 32H_2O$$

$$48e^- + 16H^+ + 16NO_3^- + 24H_2S \rightarrow 16NO + 32H_2O + 3S_8 + 48e^-$$

The equation becomes

$$16H^+(aq) + 16NO_3^-(aq) + 24H_2S(aq) \rightarrow 16NO(g) + 32H_2O(l) + 3S_8(s)$$

Step 5: Let's check the elements and the charges on each side. There are 64 hydrogen atoms, 16 nitrogen atoms, 48 oxygen atoms, and 24 sulfur atoms on each side. The charge on each side is zero, so the equation is balanced.

$$16H^+(aq) + 16NO_3^-(aq) + 24H_2S(aq) \rightarrow 16NO(g) + 32H_2O(l) + 3S_8(s)$$

elements	64 H	16 N	48 O	24 S	→	64 H	16 N	48 O	24 S
charge		0			→			0	

b. Step 1: Write the equations for the oxidation and reduction half-reactions.

$$H_5IO_6(aq) + I^-(aq) \longrightarrow I_2(s)$$

1+ each 7+ 2− each 1− 0 each

Iodine in H_5IO_6 gains seven electrons to become I^-. Iodine is reduced. The reduction half-reaction is

$$H_5IO_6 \rightarrow I_2 \quad \text{reduction half-reaction}$$

Iodide ion loses an electron to become I_2. Iodine is oxidized. The oxidation half-reaction is

$$I^- \rightarrow I_2 \quad \text{oxidation half-reaction}$$

In this oxidation–reduction reaction, iodine is both the species oxidized and the species reduced.

Step 2a: Let's balance all the elements except oxygen and hydrogen. For the reduction half-reaction there is one iodine atom on the left and two on the right. Change the coefficient of H_5IO_6 from one to two so that the number of iodine atoms is the same on each side.

$$2H_5IO_6 \rightarrow I_2$$

For the oxidation half-reaction, put a coefficient of two on the left side to balance the two iodine atoms on the right.

$$2I^- \rightarrow I_2$$

Step 2b: There are 12 oxygen atoms on the left side of the reduction half-reaction, so put 12 molecules of H_2O on the right to balance the oxygen atoms.

$$2H_5IO_6 \rightarrow I_2 + 12H_2O$$

The reduction half-reaction contains no oxygen atoms.

Step 2c: There are 24 hydrogen atoms on the right side of the reduction half-reaction, but only ten on the left. Add 14 hydrogen ions to the left side.

$$14H^+ + 2H_5IO_6 \rightarrow I_2 + 12H_2O$$

The oxidation half-reaction contains no hydrogen atoms.

Step 2d: To balance the 14+ charge on the left side of the reduction half-reaction, add 14 electrons to the left side.

$$14H^+ + 2H_5IO_6 \rightarrow I_2 + 12H_2O$$

charge $\quad 14+ \rightarrow 0$

$$14e^- + 14H^+ + 2H_5IO_6 \rightarrow I_2 + 12H_2O$$

The reduction half-reaction is now balanced.

To balance the 2− charge on the left side of the oxidation half-reaction, add two electrons to the right side.

$$2I^- \rightarrow I_2$$

charge $\quad 2- \rightarrow 0$

$$2I^- \rightarrow I_2 + 2e^-$$

Step 3: Because the reduction half-reaction transfers 14 electrons, and the oxidation half-reaction transfers two electrons, multiply the oxidation half-reaction by seven.

$$7(2I^- \rightarrow I_2 + 2e^-)$$

$$14I^- \rightarrow 7I_2 + 14e^-$$

Step 4: Now add the half-reactions together.

$$14e^- + 14H^+ + 2H_5IO_6 \rightarrow I_2 + 12H_2O$$

$$14I^- \rightarrow 7I_2 + 14e^-$$

$$14e^- + 14H^+ + 2H_5IO_6 + 14I^- \rightarrow 8I_2 + 12H_2O + 14e^-$$

The electrons on both sides cancel, and the number of iodine molecules can be combined to simplify the equation.

$$14H^+ + 2H_5IO_6 + 14I^- \rightarrow 8I_2 + 12H_2O$$

Step 5: Let's check the elements and charges on each side. There are 24 hydrogen atoms, 16 iodine atoms and 12 oxygen atoms on each side. The charge on both sides is zero, so the equation is balanced.

	$14H^+ + 2H_5IO_6 + 14I^-$			$\rightarrow$	$8I_2 + 12H_2O$		
elements	24 H	16 I	12 O	$\rightarrow$	48 H	16 I	12 O
charge		0		$\rightarrow$		0	

c. Step 1: Write equations for the oxidation and reduction half-reactions.

$$Cr_2O_7^{2-}(aq) + Sn^{2+}(aq) \longrightarrow Sn^{4+}(aq) + Cr^{3+}(aq)$$

6+ each 2– each 2+ 4+ 3+

Chromium in $Cr_2O_7^{2-}$ gains three electrons to become Cr^{3+}. Chromium is reduced. The reduction half-reaction is

$$Cr_2O_7^{2-} \rightarrow Cr^{3+} \quad \text{reduction half-reaction}$$

The Sn^{2+} ion loses two electrons to become Sn^{4+}. Tin is oxidized. The oxidation half-reaction is

$$Sn^{2+} \rightarrow Sn^{4+} \quad \text{oxidation half-reaction}$$

Step 2a: The reduction half-reaction has two chromium atoms on the left and one on the right, so change the coefficient of Cr^{3+} to two.

$$Cr_2O_7^{2-} \rightarrow 2Cr^{3+}$$

The oxidation half-reaction has one tin on each side. The coefficients need no adjustment.

Step 2b: In the reduction half-reaction there are seven oxygen atoms on the left and none on the right, so add seven molecules of H_2O to the right to balance the oxygens.

$$Cr_2O_7^{2-} \rightarrow 2Cr^{3+} + 7H_2O$$

The oxidation half-reaction contains no oxygen atoms.

Step 2c: The reduction half-reaction has 14 hydrogens on the right and none on the left, so add 14 hydrogen ions to the left side.

$$14H^+ + Cr_2O_7^{2-} \rightarrow 2Cr^{3+} + 7H_2O$$

The oxidation half-reaction needs no adjustments for hydrogens.

Step 2d: To balance the 6+ charge on the left side of the reduction half-reaction, add six electrons to the left side.

$$14H^+ + Cr_2O_7^{2-} \rightarrow 2Cr^{3+} + 7H_2O$$

charge 12+ → 6+

$$6e^- + 14H^+ + Cr_2O_7^{2-} \rightarrow 2Cr^{3+} + 7H_2O$$

The oxidation half-reaction has a 2+ charge on the left and 4+ charge on the right, so add two electrons to the right to balance the charge.

$$Sn^{2+} \rightarrow Sn^{4+}$$

charge 2+ → 6+

$$Sn^{2+} \rightarrow Sn^{4+} + 6e^-$$

Step 3: In the reduction half-reaction six electrons are gained, and in the oxidation half-reaction two electrons are lost. Multiply the oxidation half-reaction times three to equalize the number of electrons transferred.

$$3(Sn^{2+} \rightarrow Sn^{4+} + 2e^-)$$

$$3Sn^{2+} \rightarrow 3Sn^{4+} + 6e^-$$

Step 4: Now add the two half-reactions together.

$$6e^- + 14H^+ + Cr_2O_7^{2-} \rightarrow 2Cr^{3+} + 7H_2O$$

$$3Sn^{2+} \rightarrow 3Sn^{4+} + 6e^-$$

$$6e^- + 14H^+ + Cr_2O_7^{2-} + 3Sn^{2+} \rightarrow 2Cr^{3+} + 7H_2O + 3Sn^{4+} + 6e^-$$

The electrons cancel, and the equation becomes

$$14H^+ + Cr_2O_7^{2-} + 3Sn^{2+} \rightarrow 2Cr^{3+} + 7H_2O + 3Sn^{4+}$$

Step 5: Let's check the elements and charges on both sides. There are 14 hydrogen atoms, two chromium atoms, seven oxygen atoms and three tin atoms on each side. The charge is 18+ on both sides; the equation is balanced.

$$14H^+ + Cr_2O_7^{2-} + 3Sn^{2+} \rightarrow 2Cr^{3+} + 7H_2O + 3Sn^{4+}$$

elements 14 H 2 Cr 7 O 3 Sn → 14 H 2 Cr 7 O 3 Sn

charge 18+ → 18+

d. Step 1: Write equations for the oxidation and reduction half-reactions.

$$I_2(s) + NO_3^-(aq) \longrightarrow IO_3^-(aq) + NO_2(g)$$

0   5+   2− each   5+   2− each   4+   2− each

Each atom in molecular iodine loses five electrons to become IO_3^-. Iodine is oxidized. The oxidation half-reaction is

$$I_2 \rightarrow IO_3^- \quad \text{oxidation half-reaction}$$

Nitrogen in NO_3^- gains one electron to become NO_2. Nitrogen is reduced. The reduction half-reaction is

$$NO_3^- \rightarrow NO_2 \quad \text{reduction half-reaction}$$

Step 2a: There are two iodine atoms on the left side of the oxidation half-reaction and one on the right. Change the coefficient of IO_3^- to two.

$$I_2 \rightarrow 2IO_3^-$$

The reduction half-reaction has one nitrogen atom on each side.

Step 2b: The oxidation half reaction has six oxygen atoms on the right, so add six molecules of H_2O to the left.

$$6H_2O + I_2 \rightarrow 2IO_3^-$$

The reduction half-reaction has three oxygen atoms on the left but only two on the right, so add one molecule of H_2O to the right.

$$NO_3^- \rightarrow NO_2 + H_2O$$

Step 2c: The oxidation half-reaction has 12 hydrogens on the left, so add 12 hydrogen ions to the right.

$$6H_2O + I_2 \rightarrow 2IO_3^- + 12H^+$$

The reduction half-reaction has two hydrogen atoms on the right, so add two hydrogen ions to the left.

$$2H^+ + NO_3^- \rightarrow NO_2 + H_2O$$

Step 2d: To balance the 10+ charge on the right side of the oxidation half-reaction, add ten electrons to the right side.

$$6H_2O + I_2 \rightarrow 2IO_3^- + 12H^+$$

charge   0 → 10+

$$6H_2O + I_2 \rightarrow 2IO_3^- + 12H^+ + 10e^-$$

To balance the 1+ charge on the left side of the reduction half-reaction, add one electron to the left side.

$$2H^+ + NO_3^- \rightarrow NO_2 + H_2O$$

charge   1+ → 0

$$e^- + 2H^+ + NO_3^- \rightarrow NO_2 + H_2O$$

Step 3: In the oxidation half-reaction ten electrons are lost while one electron is gained during reduction. Multiply the reduction half-reaction by ten to equalize the electrons transferred.

$$10(e^- + 2H^+ + NO_3^- \rightarrow NO_2 + H_2O)$$

$$10e^- + 20H^+ + 10NO_3^- \rightarrow 10NO_2 + 10H_2O$$

Step 4: Now add the two half-reactions together.

$$6H_2O + I_2 \rightarrow 2IO_3^- + 12H^+ + 10e^-$$

$$10e^- + 20H^+ + 10NO_3^- \rightarrow 10NO_2 + 10H_2O$$

$$\overline{10e^- + 8H^+ + 10NO_3^- + I_2 \rightarrow 10NO_2 + 4H_2O + 2IO_3^- + 10e^-}$$

The ten electrons on each side cancel. There are 20 hydrogen ions on the left and 12 on the right, so cancel the 12 hydrogen ions on the right, and leave eight on the left. There are six water molecules on the left and ten on the right so cancel the six molecules on the left, which leaves four on the right. The equation becomes

$$8H^+(aq) + 10NO_3^-(aq) + I_2(s) \rightarrow 10NO_2(g) + 4H_2O(l) + 2IO_3^-(aq)$$

Step 5: Let's check the elements and charges on each side. There are eight hydrogen atoms, ten nitrogen atoms, 30 oxygen atoms and two iodine atoms on each side. The charge on each side is 2−, so the equation is balanced.

	$8H^+(aq) + 10NO_3^-(aq) + I_2(s)$	→	$10NO_2(g) + 4H_2O(l) + 2IO_3^-(aq)$
elements	8 H 10 N 30 O 2 I	→	8 H 10 N 30 O 2 I
charge	2−	→	2−

9. Usually redox reactions occur in a single container. If we separate the oxidation half-reaction from the reduction half-reaction but use a salt bridge to allow ions to flow, we can require the electron transfer to occur through a wire. The current produced can be used to do work.

10. During oxidation, electrons are lost, so oxidation occurs in the left container and reduction in the right container.

11. In the lead storage battery there are lead grids connected by a metal bar. Lead is oxidized to form Pb^{2+}, which combines with SO_4^{2-} from sulfuric acid (battery acid) to form solid $PbSO_4$. The substance that gains electrons and is reduced is PbO_2, which is coated on the lead grids. Pb^{4+} in PbO_2 is reduced to Pb^{2+}, which combines with SO_4^{2-} from H_2SO_4 to form $PbSO_4$. So the product of both oxidation and reduction is $PbSO_4$. The oxidation and reduction half-reactions are separated so that useful work, such as starting your car, can be accomplished.

12. Aluminum metal can be produced by electrolysis of aluminum oxide. Electrolysis is the process of forcing a current through an electrochemical cell to cause a chemical change that would not occur naturally. During electrolysis, aluminum is reduced.

$$2Al_2O_3(s) \rightarrow 4Al(s) + 3O_2(g)$$

CHAPTER 19

Radioactivity and Nuclear Energy

INTRODUCTION

Most chemical properties depend on the arrangement of electrons, and many chemical reactions involve the transfer of electrons from one atom to another. But the events and reactions described in this chapter depend on the properties of the nucleus of an atom. The best-known nuclear reactions produce energy in nuclear reactors and in nuclear explosions. You will learn about these reactions and other processes in this chapter.

CHAPTER DISCUSSION

Not all nuclei are stable. Many decay spontaneously, producing a new nucleus, and in addition, some type of nuclear particle. The nucleus of cobalt-60 is unstable. It spontaneously decays to produce nickel-60 and an electron. The reaction can be written

$$^{60}_{27}\text{Co} \rightarrow \, ^{60}_{28}\text{Ni} + \, ^{0}_{-1}\text{e}$$

When writing nuclear decay reactions, always show the atomic mass, A, and the atomic number, Z, for each element and each particle. Nuclear equations must be balanced. The sum of the atomic masses must be the same on both sides of the equation, and the sum of the atomic numbers must be the same on both sides. In the equation above, the atomic mass is 60 on the left side and 60 plus zero on the right. The atomic number is 27 on the left side and 28 minus one, or 27, on the right. Both sides are balanced.

Different types of radioactive decay are defined by the type of particle produced in the reaction. One type of decay produces an alpha particle, which is a helium nucleus. Beta particles also are often produced during nuclear decay. A beta particle is an electron. The net effect of beta production is to change a neutron to a proton. The atomic number in the new nucleus increases by one. The mass number does not change.

Sometimes gamma rays are produced during nuclear decay, usually accompanied by another particle. A gamma ray is a high-energy photon of light with no mass or atomic number.

A positron has very little mass and a positive charge. The production of a positron does not change the mass number but decreases the atomic number by one. Sometimes the nucleus can capture one of its own inner orbital electrons. This process is called electron capture. The capture of an electron decreases the atomic number by one.

Examples of each:

Alpha-particle production: $^{211}_{83}\text{Bi} \rightarrow \, ^{4}_{2}\text{He} + \, ^{207}_{81}\text{Tl}$

Beta-particle production: $^{10}_{4}\text{Be} \rightarrow \, ^{0}_{-1}\text{e} + \, ^{10}_{5}\text{B}$

Gamma-ray production: $^{226}_{88}\text{Ra} \rightarrow \, ^{222}_{86}\text{Rn} + \, ^{4}_{2}\text{He} + \, ^{0}_{0}\gamma$

Positron production: $^{15}_{8}\text{O} \rightarrow \, ^{15}_{7}\text{N} + \, ^{0}_{+1}\text{e}$

Electron capture: $^{37}_{18}\text{Ar} + \, ^{0}_{-1}\text{e} \rightarrow \, ^{37}_{17}\text{Cl}$

LEARNING REVIEW

1. In a balanced nuclear equation, which two quantities must be the same on both sides of the equation?

2. What is the atomic number and the mass number of each of the particles below?

   a. gamma ray

   b. positron

   c. alpha particle

   d. beta particle

3. Write balanced nuclear equations for the decay of the radioactive particles below.

   a. $^{226}_{86}Rn$ decays to produce an α-particle and a γ-ray.

   b. $^{70}_{31}Ga$ decays to produce a β-particle.

   c. $^{144}_{60}Nd$ decays to produce a β-particle.

4. Complete and balance these nuclear equations.

   a. $^{161}_{67}Ho + ? \rightarrow\ ^{161}_{66}Dy$

   b. $^{10}_{4}Be \rightarrow ? +\ ^{0}_{-1}e$

   c. $? +\ ^{0}_{-1}e \rightarrow\ ^{44}_{21}Sc$

   d. $^{253}_{99}Es +\ ^{4}_{2}He \rightarrow\ ^{1}_{1}H + ?$

   e. $^{59}_{29}Cu \rightarrow ? +\ ^{59}_{28}Ni$

5. Show the product formed when the nuclide below is bombarded with a smaller nuclide.

$$^{238}_{92}U +\ ^{12}_{6}C \rightarrow ? +\ ^{1}_{0}n$$

6. Two instruments for detecting radioactivity are the Geiger counter and the scintillation counter. Briefly explain how each one works.

7. In a sample of the nuclides below, which would exhibit the highest number of decay events during a fixed period of time?

Name	Half-life
a. potassium-42	12.4 hours
b. hydrogen-3	12.5 years
c. plutonium-239	2.44×10^4 years

8. If a sample of 5.0×10^{20} iodine-131 atoms with a half-life of eight days is allowed to decay for 48 days, how many iodine-131 atoms will remain?

9. A wooden post from an ancient village has 25% of the carbon-14 found in living trees. How old is the wooden post? The half-life of carbon-14 is 5730 years.

10. Why do you think that most nuclides used in medicine as radiotracers have short half-lives?

11. What safety features would prevent a nuclear explosion in case of a serious malfunction of a nuclear reactor?
12. Why do you think that the fusion process would supplant fission if the technology were available?
13. What differences exist between genetic and somatic damage caused by radioactivity?
14. Why is the ionizing ability of a radiation source important in determining the biological effects of radiation?

ANSWERS TO LEARNING REVIEW

1. The sum of the atomic numbers (*Z*) and the sum of the mass numbers (*A*) must be the same on both sides of a nuclear equation.
2.
   a. A gamma ray has a mass number of zero and an atomic number of zero.
   b. A positron has a mass number of zero and an atomic number of 1+.
   c. An alpha particle has a mass number of four and an atomic number of 2+.
   d. A beta particle has a mass number of zero and an atomic number of 1−.
3.
   a. When Rn-226 decays to produce an alpha particle and a gamma particle, the mass number of the new nuclide is decreased by four to 222. The atomic number decreases by two to 84. The new nuclide would have a mass number of 222 and an atomic number of 84. The element with atomic number of 84 is polonium, so the new nuclide is Po-222.

$$^{226}_{86}\text{Rn} \rightarrow {}^{4}_{2}\text{He} + {}^{0}_{0}\gamma + {}^{222}_{84}\text{Po}$$

   b. When Ga-70 decays to produce a beta particle, the mass number of the new nuclide does not change. The atomic number increases by one to 32. The new nuclide would have a mass number of 70, and an atomic number of 32. The element with an atomic number of 32 is germanium, so the new nuclide is Ge-70.

$$^{70}_{31}\text{Ga} \rightarrow {}^{0}_{-1}\text{e} + {}^{70}_{32}\text{Ge}$$

   c. When 60 Nd-144 decays to produce a beta particle, the mass number does not change. The atomic number increases by one to 61. The new nuclide would have a mass number of 144 and an atomic number of 61. The element with an atomic number of 61 is promethium, so the new nuclide is Pm-144.

$$^{144}_{60}\text{Nd} \rightarrow {}^{0}_{-1}\text{e} + {}^{144}_{61}\text{Pm}$$

4.
   a. This problem provides the nuclides before and after decay and asks for the identity of an unknown decay particle. Because the mass number does not change on either side, the mass number of the particle is zero. The atomic number decreases by one on the right side, so the atomic number of the unknown particle is 1− so that the sum of atomic numbers on each side is the same. The unknown particle has a mass number of zero and an atomic number of −1. It is a β-particle. This is an example of electron capture.

$$^{161}_{67}\text{Ho} + {}^{0}_{-1}\text{e} \rightarrow {}^{161}_{66}\text{Dy}$$

b. This problem provides a nuclide on the left that decays to an unknown nuclide and a β-particle. Because the mass number of the β-particle is zero, the mass number of the unknown nuclide must be ten to balance the left side. The atomic number increases by one to become five to balance the four on the left side. The element with an atomic number of five is boron, so the nuclide is B-10.

$$^{10}_{4}Be \rightarrow\ ^{10}_{5}B +\ ^{0}_{-1}e$$

c. This problem provides the identity of a particle that combines with an unknown nuclide to produce the nuclide Sc-44. The β-particle has a mass number of zero, so the mass number of the unknown nuclide must be 44. The β-particle has an atomic number of 1-, so the atomic number of the unknown nuclide must be 22 so that the sum of the atomic numbers is the same on each side. The element with atomic number 22 is titanium, so the unknown nuclide is Ti-44.

$$^{44}_{22}Ti +\ ^{0}_{-1}e \rightarrow\ ^{44}_{21}Sc$$

d. This problem provides a nuclide that reacts with an alpha particle to produce a proton and an unknown nuclide. The total mass number on the left side is 257. On the right, the proton has a mass number of one, so the unknown nuclide must have a mass number of 256 so that both sides are balanced. The total atomic number on the left side is 101. On the right, the proton has an atomic number of one, so that the atomic number of the unknown nuclide must be 100. The element with an atomic number of 100 is fermium, so the nuclide is Fm-256.

$$^{253}_{99}Es +\ ^{4}_{2}He \rightarrow\ ^{1}_{1}H +\ ^{256}_{100}Fm$$

e. This problem provides a nuclide that decays to an unknown particle and to a nuclide of nickel. The mass number on both sides is 59, so the mass number of the unknown particle must be zero. The atomic number of the nuclide on the left is 29, and the atomic number of the nuclide on the right is 28. The unknown particle has an atomic number of 1+. The particle with a mass number of zero and an atomic number of 1+ is a positron.

$$^{59}_{29}Cu \rightarrow\ ^{0}_{+1}e +\ ^{59}_{28}Ni$$

5. The problem provides a nuclide of uranium that is bombarded with a smaller carbon nucleus to produce an unknown nuclide and four neutrons. The total mass number on the left is 238 plus 12, which is 250. Each neutron on the right has a mass number of one, so the total mass number of the neutrons is four. The mass number of the new nuclide is 246. The total atomic number on the left is 92 plus six, which is 98. Each of the four neutrons on the right has an atomic number of zero, so the atomic number of the new nuclide is 98. The element with an atomic number of 98 is californium, Cf. The new nuclide is Cf-246.

$$^{238}_{92}U +\ ^{12}_{6}C \rightarrow\ ^{246}_{98}Cf + 4\,^{1}_{0}n$$

6. The Geiger-Müller counter, or Geiger counter, has a probe that is placed close to the source of radioactivity. The probe contains atoms of argon gas that lose an electron when hit by a high-speed subatomic particle. The argon cation and accompanying electron produce a momentary pulse of electrical current that is detected by the Geiger counter. The amount of radioactive material is directly related to the number of pulses detected.

Radioactivity can also be detected with a scintillation counter. High-speed decay particles collide with a substance inside the scintillation counter such as sodium iodide. The sodium iodide emits a flash of light when struck. Each flash of light is counted, and the number of flashes is directly related to the amount of radioactivity.

7. The half-life of potassium-42 is 12.4 hours, which means that 50% of a sample of potassium-42 would decay in 12.4 hours. Plutonium-239 has a half-life of 24,400 years, which means it would take 24,400 years for 50% of a plutonium-239 sample to decay. The shorter the half-life, the quicker a nuclide decays, so the nuclide with the smallest half-life produces the most decay events over time. Of the three nuclides, potassium-42 would produce the most decay events in any fixed amount of time.

8. Because the half-life of iodine-131 is eight days, the number of iodine-131 atoms in any sample will decrease by 50% after eight days. So, after eight days there will be $(5.0 \times 10^{20})/2 = 2.5 \times 10^{20}$ iodine-131 atoms left. After another eight days (for a total of sixteen days) there would be $(2.5 \times 10^{20})/2 = 1.3 \times 10^{20}$ iodine-131 atoms left. After another eight days (for a total of twenty-four days) there would be $(1.3 \times 10^{20})/2 = 6.3 \text{ x } 10^{19}$ iodine-131 atoms left. After three more eight-day periods (for a total of 48 days) there would be 7.8×10^{18} iodine-131 atoms left.

9. A piece of wood that contains 25% of the carbon-14 found in freshly cut wood has undergone two half-life decays. The first half-life would decrease the carbon-14 from 100% to 50%, and the second half-life would decrease the carbon-14 content from 50% to 25%. So, a piece of wood that has undergone two half-life decays would be two times 5730, or 11,460, years old.

10. Using any radiotracer inside the human body poses some risk of damage by the high-speed decay particles. Radiotracers with a short half-life will rapidly decay and produce many decay particles in a short period of time. Doctors can use small amounts of radiotracer and still detect their presence because the numbers of decay particles are high at first. Because the half-life is short, most of the radiotracer usually decays quickly.

11. Nuclear reactors have many safety features, including control rods made of substances that absorb neutrons. The control rods can be raised or lowered among the fuel rods to control how fast the nuclear reaction occurs. If a serious problem occurs with the reactor, the control rods automatically lower into the core so that the fission process slows down. The amount of fissionable fuel present in any nuclear reactor is below the critical mass, so that even in the worst possible case a nuclear explosion would not occur.

12. Fusion would quickly supplant fission because fuel for fusion is readily available in sea water. Fusion reactors would produce helium as an end product and not the wide variety of radionuclides produced from fission. Safe disposal of the nuclear waste from fission is a concern that does not occur with fusion reactors.

13. Somatic damage is the damage done directly to the tissues of the organism. Somatic damage usually occurs soon after exposure to the radiation source. Genetic damage is the kind of damage done to the reproductive machinery of the human body. Genetic damage occurs at the time of exposure but may not show up until the birth of offspring.

14. When biomolecules are ionized by a radiation source, they no longer perform their functions in the body.

CHAPTER 20

Organic Chemistry

INTRODUCTION

Organic chemistry is devoted to the study of compounds and reactions of the element carbon. No other element forms as many different compounds as carbon. The compounds range from the simple molecule methane, which we burn as fuel, to the complex molecules that carry genetic information. This chapter will help you learn the language of organic chemistry by introducing you to how carbon-containing molecules are formed, how they function, and how they are named.

CHAPTER DISCUSSION

A carbon atom can form a maximum of four covalent bonds with other atoms. The VSEPR model tells us that electron pairs try to spread out as far as possible. The shape assumed by a carbon atom bonded to four atoms is a tetrahedron. Carbon can bond to fewer than four other atoms when it forms a double or triple bond. When carbon forms a double bond with an atom, two electron pairs are used to make the double bond. When carbon forms a triple bond, three electron pairs are used to form the triple bond. Carbons with either double or triple bonds do not have a tetrahedral shape.

Hydrocarbons are compounds composed entirely of carbon and hydrogen. Hydrocarbons are either saturated or unsaturated. A saturated hydrocarbon is one in which all the carbons are bonded to four atoms. An unsaturated hydrocarbon has at least one carbon atom bonded to less than four atoms, that is, it has at least one double or triple bond. Alkanes are saturated hydrocarbons. Alkenes are hydrocarbons that contain at least one carbon-carbon double bond. Alkynes are hydrocarbons that contain at least one carbon-carbon triple bond. The rules for naming all of these are listed in your textbook. You can practice this in the Learning Review section, along with looking at a discussion of the solutions.

In addition to alkanes, alkenes, and alkynes, you should know how to name organic compounds that are aromatic (and substituted) or that contain functional groups. These include alcohols (primary, secondary, and tertiary), aldehydes, ketones, carboxylic acids, and esters. Naming these compounds is quite systematic and presented quite clearly in the text. There are also many practice problems in the Learning Review section of this Study Guide, along with their solutions.

LEARNING REVIEW

1. Match the term below with the correct definition.

a. hydrocarbon	contains one or more double or triple bonds
b. alkane	an unbranched molecule
c. normal	contains only carbon-carbon single bonds
d. unsaturated	composed of carbon and hydrogen

2. What is the shape of a carbon tetrachloride molecule, CCl_4?

3. Write the condensed formula for an unbranched alkane with five carbons.

4. A branched alkane with the formula below has seven carbons. Can this alkane be represented by the general formula C_nH_{2n+2}?

```
CH3—CH2—CH2—CH—CH3
             |
            CH2
             |
            CH3
```

5. What are structural isomers?

6. Write all structural isomers of hexane, C_6H_{14}.

7. Name the alkanes below.

a.
```
CH3—CH—CH—CH2—CH—CH3
    |   |       |
   CH3 CH3     CH3
```

b.
```
CH3—CH2—CH—CH2—CH3
         |
         CH—CH3
         |
        CH3
```

c.
```
CH3—CH—CH—CH3
    |   |
   CH2 CH3
    |
   CH3
```

d.
```
                   CH3
                    |
                   CH2
                    |
CH3—CH2—CH—CH—CH—CH2—CH3
         |   |
        CH2 CH3
         |
        CH3
```

e.
```
     CH3
      |
H3C—C—CH3
      |
     CH2
```

8. Are the molecules a and b structural isomers of each other?

a.
```
CH3—CH—CH—CH3
    |   |
   CH3 CH2—CH3
```

b.
```
    CH3
     |
CH3—CH  CH3
     |   |
    CH2—CH
         |
        CH3
```

9. Write structural formulas for the molecules below.

   a. 2-methyl-4-*sec*-butyloctane

   b. 2-methylpropane

   c. 3-ethylpentane

   d. 2,2,4-trimethylhexane

10. Each of the alkane names below is incorrect. Write the structural formula for each molecule; then name it correctly.

    a. 2-ethylbutane

    b. 1-methylpentane

    c. 4,4-dimethylhexane

    d. 2-ethyl-2-methylheptane

11. Petroleum itself is not very useful. It must be separated into fractions to be useful. How is petroleum separated into the fractions that have useful properties?

12. When chlorine gas reacts with pentane, one of the hydrogens is substituted by a chlorine atom. Several structural isomers are possible depending on which hydrogen atom is substituted. Show all possible structural isomers that could be produced when chlorine reacts with pentane in the presence of ultraviolet light.

13. Write the structures for the products of the reactions below.

    a. 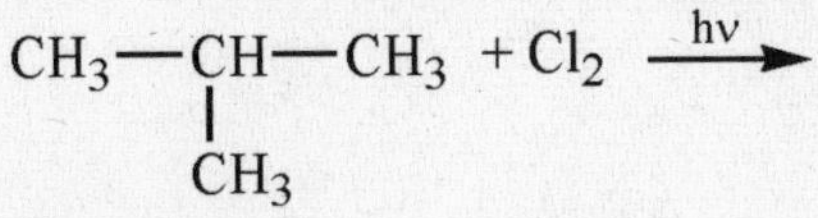 $CH_3-\underset{\displaystyle CH_3}{\underset{|}{CH}}-CH_3 + Cl_2 \xrightarrow{h\nu}$

    b. $CH_4 + O_2 \longrightarrow$

    c. $CH_3-CH_2-CH_3 \xrightarrow[500°C]{Cr_2O_3}$

14. Name each of the molecules below.

    a. 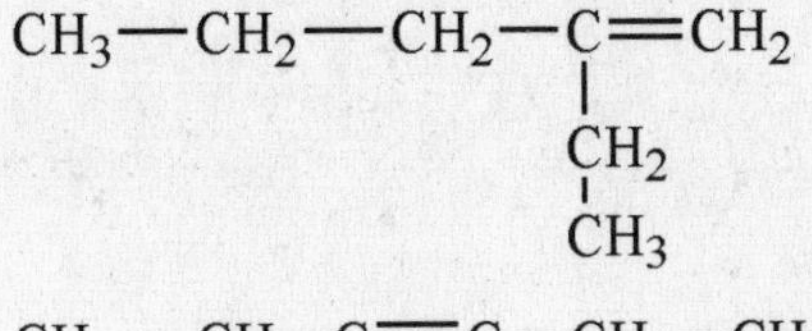 $CH_3-CH_2-CH_2-\underset{\displaystyle \underset{\displaystyle CH_3}{\underset{|}{CH_2}}}{\underset{|}{C}}=CH_2$

    b. $CH_3-\underset{\displaystyle CH_3}{\underset{|}{CH}}-C\equiv C-CH_2-\underset{\displaystyle CH_3}{\underset{|}{CH}}-CH_3$

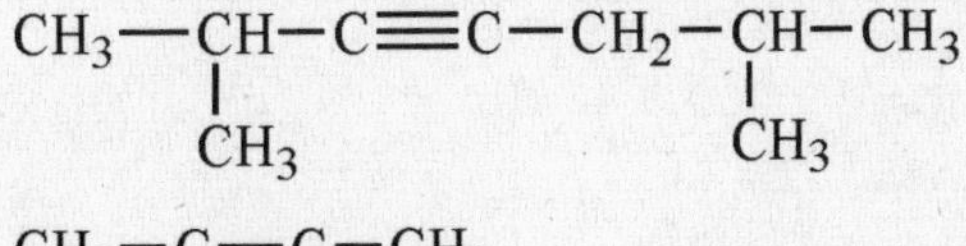

    c. $CH_3-\underset{\displaystyle CH_3}{\underset{|}{C}}=\underset{\displaystyle CH_3}{\underset{|}{C}}-CH_3$

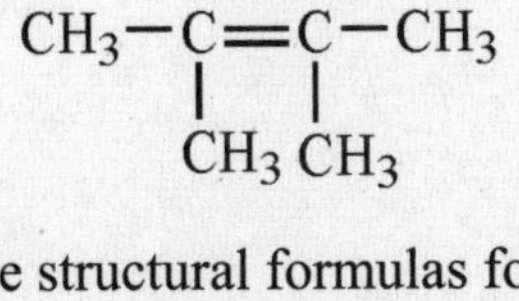

15. Write structural formulas for the products of the reactions below.

    a. 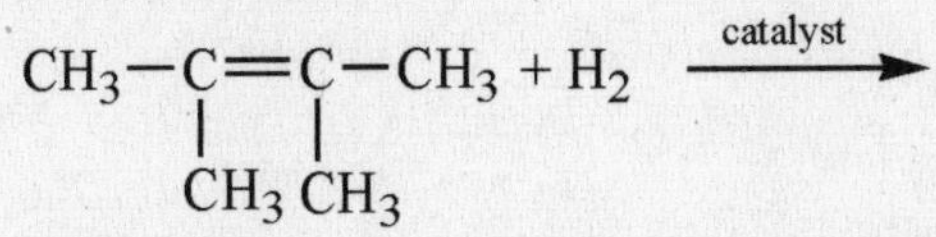 $CH_3-\underset{\displaystyle CH_3}{\underset{|}{C}}=\underset{\displaystyle CH_3}{\underset{|}{C}}-CH_3 + H_2 \xrightarrow{\text{catalyst}}$

    b. $CH_3-CH_2-CH_2-CH=CH_2 + Cl_2 \longrightarrow$

c. 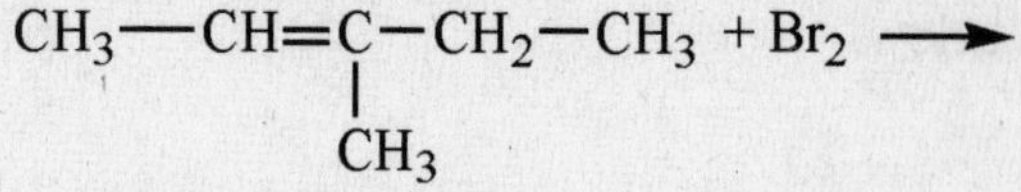

16. Name the aromatic hydrocarbons below.

a. 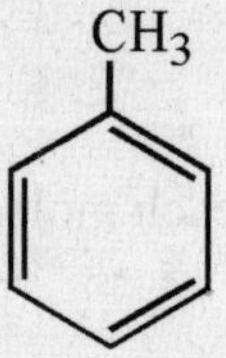

b. 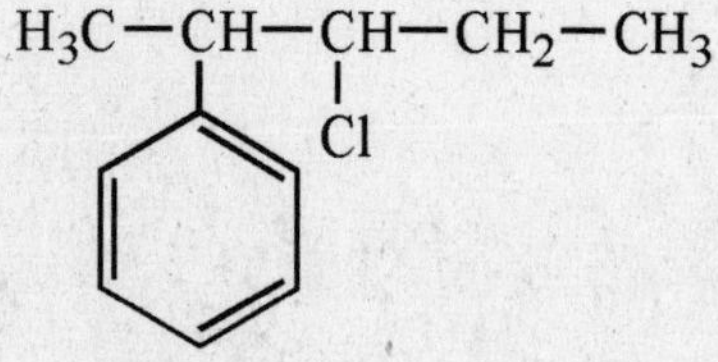

c. 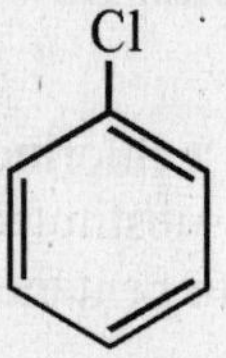

d. 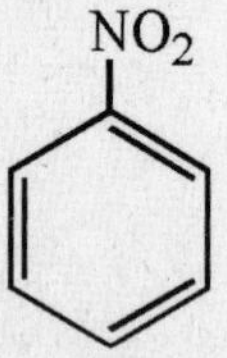

17. Name the di-substituted aromatic hydrocarbons below.

a. 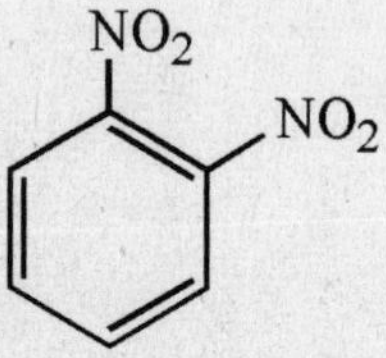

b. 

c. 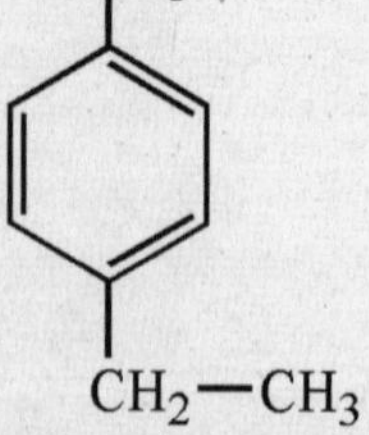

CH_3

Br

d.

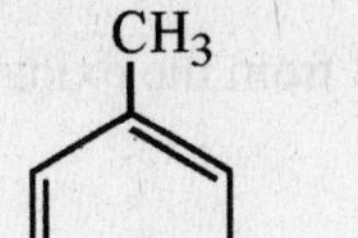

18. Name each of the alcohols below, and decide whether they are primary, secondary, or tertiary alcohols.

a.

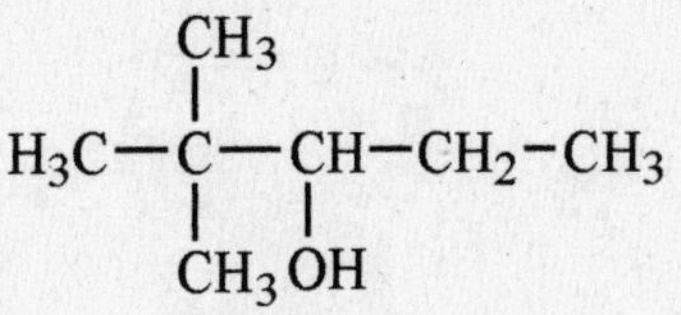

$$\begin{array}{l} \quad\ \ CH_3 \\ \quad\ \ | \\ H_3C-C-CH-CH_2-CH_3 \\ \quad\ \ |\quad\ \ | \\ \quad\ CH_3\ OH \end{array}$$

b.

$$\begin{array}{l} \qquad\quad CH_3 \\ \qquad\quad | \\ CH_3-C-OH \\ \qquad\quad | \\ \qquad\quad CH_2 \\ \qquad\quad | \\ \qquad\quad CH_3 \end{array}$$

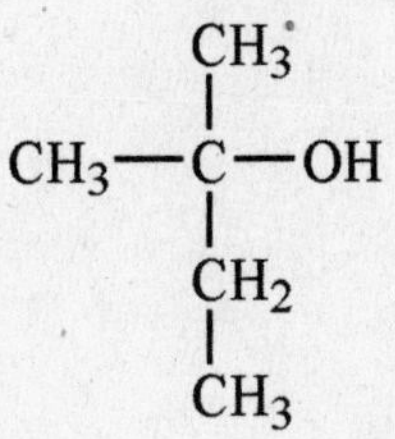

c.

$$\begin{array}{l} \qquad\qquad\qquad CH_3 \\ \qquad\qquad\qquad | \\ CH_3-CH_2-C-CH-CH_2-CH_2-OH \\ \qquad\qquad\qquad |\quad\ \ | \\ \qquad\qquad\quad\ Br\ \ CH_3 \end{array}$$

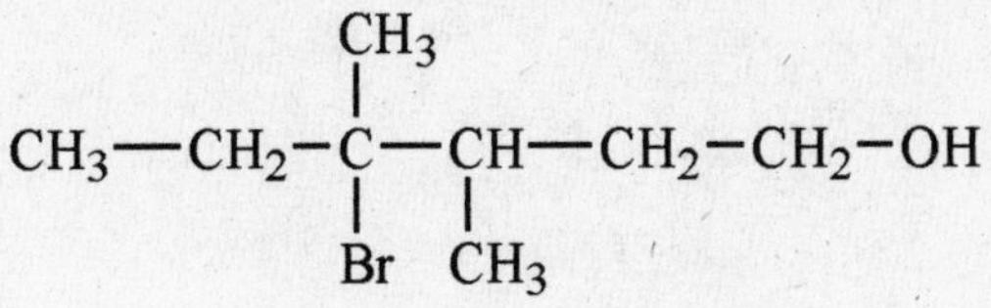

d.

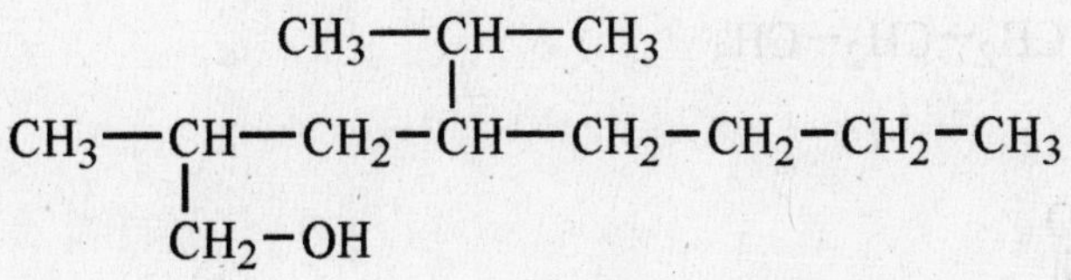

$$\begin{array}{l} \qquad\qquad\qquad CH_3-CH-CH_3 \\ \qquad\qquad\qquad\qquad\ \ | \\ CH_3-CH-CH_2-CH-CH_2-CH_2-CH_2-CH_3 \\ \qquad\quad\ | \\ \qquad\quad CH_2-OH \end{array}$$

e. CH_3-OH

f.

$$\begin{array}{l} CH_3-CH_2-CH-CH_2-CH_3 \\ \qquad\qquad\quad\ | \\ \qquad\qquad CH_2-CH_2-CH-CH_2-CH-CH_3 \\ \qquad\qquad\qquad\qquad\quad\ |\qquad\qquad\ | \\ \qquad\qquad\qquad\qquad\ CH_2\qquad\quad OH \\ \qquad\qquad\qquad\qquad\quad\ | \\ \qquad\qquad\qquad\qquad\ CH_3 \end{array}$$

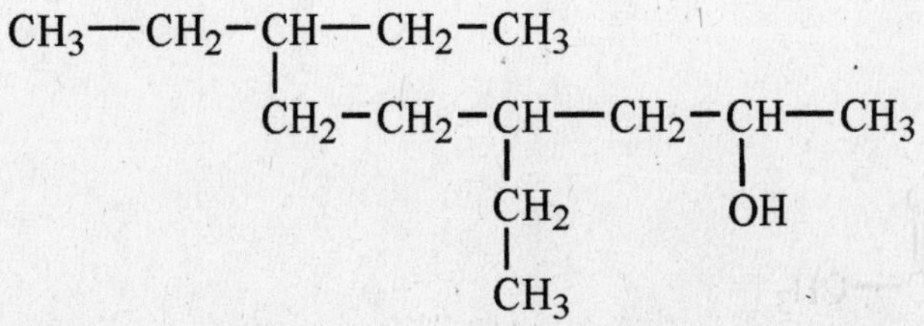

19. Match the alcohols below with the appropriate description.

a. methanol	used in antifreeze
b. phenol	found in alcoholic beverages
c. ethanol	commonly known as wood alcohol
d. ethylene glycol	used in the production of plastics

20. Show the structures of the aldehydes and ketones that would be produced from the oxidation of the following alcohols.

a. (benzene ring)$-CH_2-OH$ $\xrightarrow{\text{oxidation}}$

b. $CH_3-\underset{\displaystyle CH_3}{\underset{|}{CH}}-CH_2-OH \xrightarrow{\text{oxidation}}$

c. $CH_3-CH_2-\underset{\displaystyle OH}{\underset{|}{CH}}-CH_2-CH_3 \xrightarrow{\text{oxidation}}$

21. Name the aldehydes and ketones below.

a. $CH_3-\overset{\displaystyle CH_3}{\overset{|}{\underset{\displaystyle CH_3}{\underset{|}{C}}}}-\underset{\displaystyle CH_3}{\underset{|}{CH}}-C(=O)H$

b. $CH_3-\overset{\displaystyle O}{\overset{\|}{C}}-CH_2-\overset{\displaystyle CH_2-CH_3}{\overset{|}{CH}}-CH_2-CH_3$

c. $H(O=)C-\underset{\displaystyle Br}{\underset{|}{CH}}-CH_2-CH_2-CH_2-CH_2-CH_3$

d. $CH_3-CH_2-\underset{\displaystyle CH_3}{\underset{|}{CH}}-\underset{\displaystyle CH_2-CH_3}{\underset{|}{CH}}-\overset{\displaystyle O}{\overset{\|}{C}}-CH_3$

e. (benzene ring)$-CH_2-\underset{\displaystyle CH_3}{\underset{|}{CH}}-\overset{\displaystyle O}{\overset{\|}{C}}-CH_3$

f. $CH_3-\overset{\displaystyle CH_3}{\overset{|}{\underset{\displaystyle C(=O)H}{\underset{|}{C}}}}-\overset{\displaystyle Cl}{\overset{|}{\underset{\displaystyle Cl}{\underset{|}{C}}}}-CH_2-CH_3$

22. Match the structure with the functional group name.

a. 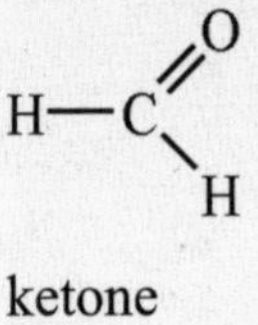 $H-C(=O)-H$

ketone

b. 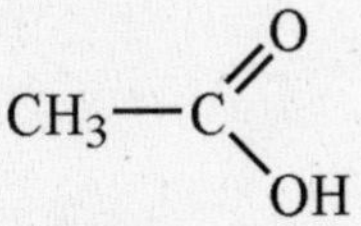 $CH_3-C(=O)-OH$

ester

c. CH_3-CH_2-OH

aldehyde

d. 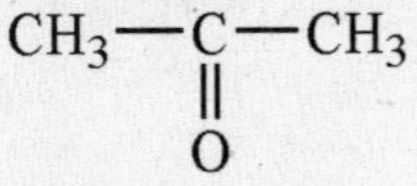 $CH_3-C(=O)-CH_3$

carboxylic acid

e. 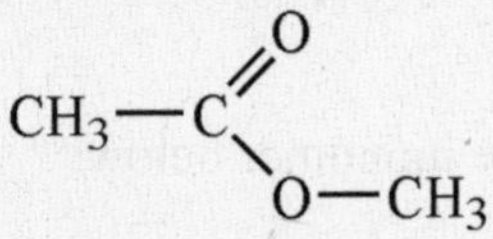 $CH_3-C(=O)-O-CH_3$

alcohol

23. Show the structure of the carboxylic acid that is produced by oxidizing primary alcohols.

a. 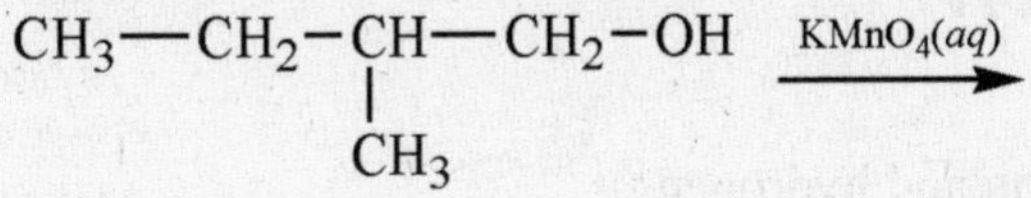 $CH_3-CH_2-CH(CH_3)-CH_2-OH \xrightarrow{KMnO_4(aq)}$

b. $CH_3-CH_2-CH_2-CH_2-CH_2-OH \xrightarrow{KMnO_4(aq)}$

24.

a. Show the structure of the ester formed when benzoic acid reacts with CH_3OH.

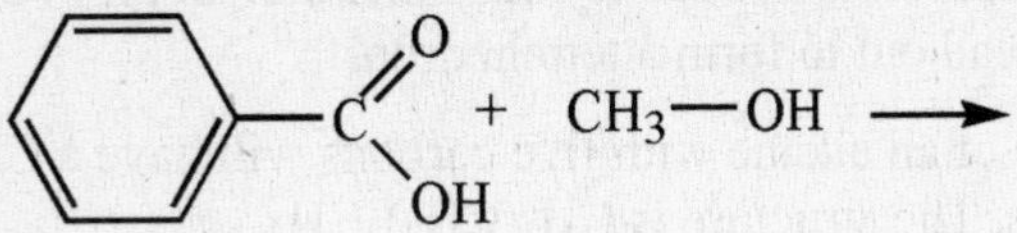

$C_6H_5-C(=O)-OH + CH_3-OH \longrightarrow$

b. What is the name of the ester?

25. Name the carboxylic acids below.

a. 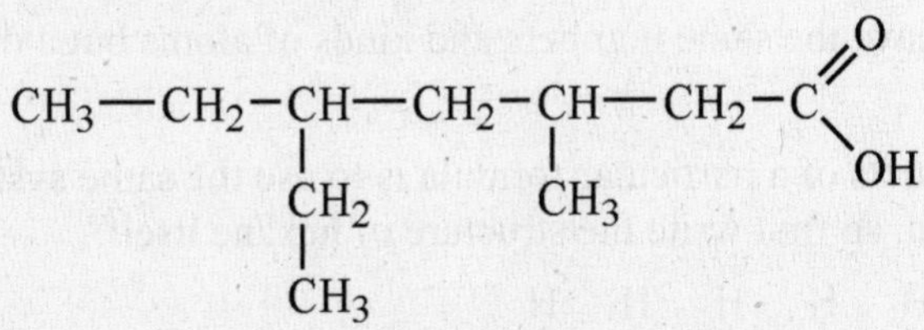 $CH_3-CH_2-CH(CH_2CH_3)-CH_2-CH(CH_3)-CH_2-C(=O)-OH$

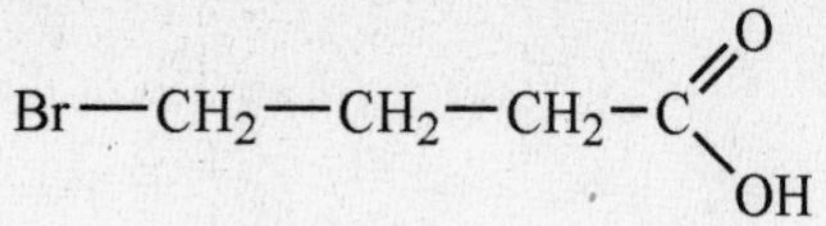

b. $Br-CH_2-CH_2-CH_2-C(=O)OH$

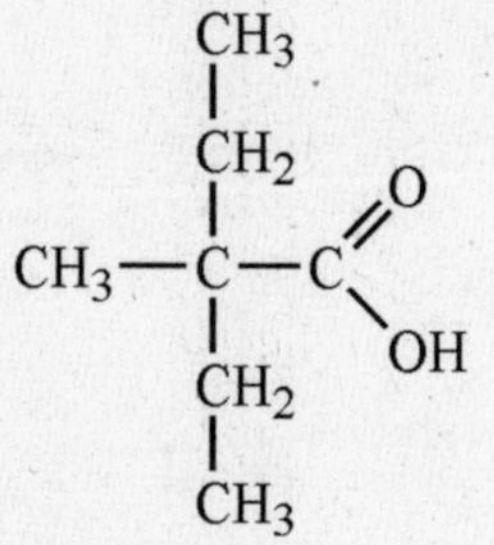

c. $CH_3-C(CH_2CH_3)(CH_2CH_3)-C(=O)OH$

26. Show the structure of the carboxylic acid and the alcohol that reacted to make the esters below.

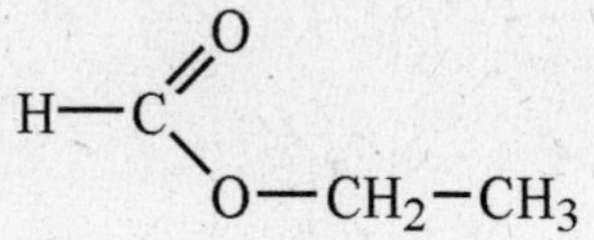

a. $H-C(=O)-O-CH_2-CH_3$

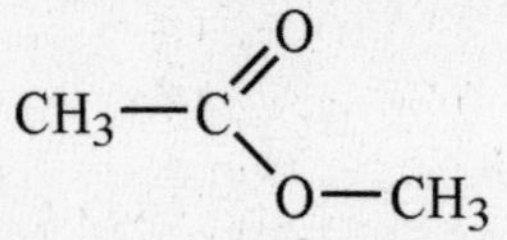

b. $CH_3-C(=O)-O-CH_3$

27. Show the structure of the polymer that would be produced from the monomer below.

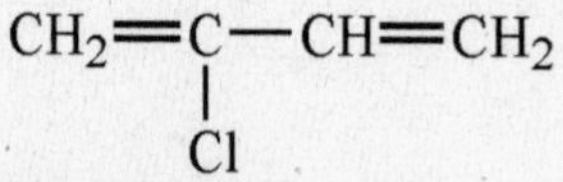

$$CH_2{=}C(Cl)-CH{=}CH_2$$

ANSWERS TO LEARNING REVIEW

1. The correct matches are shown below.

a.	hydrocarbon	composed of carbon and hydrogen
b.	alkane	contains only carbon-carbon single bonds
c.	normal	an unbranched molecule
d.	unsaturated	contains one or more double or triple bonds

2. In carbon tetrachloride, carbon is bonded to four other atoms. From the VSEPR model, four pairs of bonding electrons spread out to form a tetrahedron.

3. The condensed structure of an alkane with five carbons will have a CH_3 on each end and three CH_2, or methylene, units. The structure is $CH_3(CH_2)_3CH_3$.

4. This alkane has seven carbons and 16 hydrogens, so it can be represented by the general formula C_nH_{2n+2}. Both straight-chain and normal alkanes are represented by the general formula C_nH_{2n+2}.

5. Structural isomers are molecules that have the same numbers and kinds of atoms but a different arrangement of bonds.

6. One way to write all the structural isomers of a particular formula is to use the same system each time. Hexane has six carbons in a chain, so first write the structure of hexane itself.

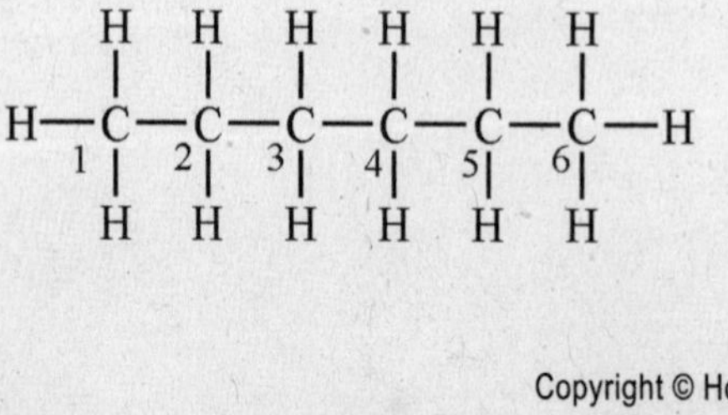

$$H-\underset{1}{C}H_2-\underset{2}{C}H_2-\underset{3}{C}H_2-\underset{4}{C}H_2-\underset{5}{C}H_2-\underset{6}{C}H_2-H$$

Then remove a $-CH_3$ from the end of hexane, which leaves five carbons in the chain. Remove an $-H$ from successive carbons on the chain, and replace it with the $-CH_3$. If we remove an $-H$ from the end carbon and replace with the $-CH_3$, we have not made a structural isomer because the new molecule is just like hexane.

```
   H   H   H   H   H             H
   |   |   |   |   |             |
H—C—C—C—C—C—          —C—H
  1|  2|  3|  4|  5|             |
   H   H   H   H   H             H
```

Put the $-CH_3$ on carbon number two. The $-CH_3$ is now located on the second carbon from the end. This is a new structural isomer.

```
   H   H   H   H   H
   |   |   |   |   |
H—C—C—C—C—C—H
  1|  2|  3|  4|  5|
   H   |   H   H   H
       |
    H—C—H
       |
       H
```

Now move the $-CH_3$ to carbon number three.

```
   H   H   H   H   H
   |   |   |   |   |
H—C—C—C—C—C—H
  1|  2|  3|  4|  5|
   H   H   |   H   H
           |
        H—C—H
           |
           H
```

The $-CH_3$ is now located on the middle carbon, the third carbon from the end. We can move the $-CH_3$ to carbon number four, but this does not create a molecule with a different order. The carbons have the same order as when the $-CH_3$ is on the second carbon from the left. Both structures have $-CH_3$ on the second carbon from the end carbon.

```
   H   H   H   H   H                   H   H   H   H   H
   |   |   |   |   |                   |   |   |   |   |
H—C—C—C—C—C—H               H—C—C—C—C—C—H
  1|  2|  3|  4|  5|                  5|  4|  3|  2|  1|
   H   |   H   H   H                   H   H   H   |   H
       |                                           |
    H—C—H                                       H—C—H
       |                                           |
       H              same as                      H
```

We have exhausted the structural isomers that can be made by moving one $-CH_3$ from one location to another. Let's now remove another carbon from the chain and see what isomers can be made from a chain with four carbons and two $-CH_3$ groups.

```
   H   H   H   H        H        H
   |   |   |   |        |        |
H—C—C—C—C—    —C—    —C—H
  1|  2|  3|  4|       5|       6|
   H   H   H   H        H        H
```

Substitute one $-CH_3$ on the second carbon and one on the third carbon.

```
       H   H              H   H
       |   |              |   |
    H—C—C————————C—C—H
      1|  2|             3|  4|
       H   |              |   H
           |              |
        H—C—H        H—C—H
           |              |
           H              H
```

Now substitute both $-CH_3$ groups on the second carbon.

```
             H
             |
          H—C—H
             |
       H     |    H    H
       |     |    |    |
    H—C—C—C—C—H
      1|   2|   3|   4|
       H     |    H    H
             |
          H—C—H
             |
             H
```

If we put both $-CH_3$ groups on the third carbon, we have not created a structure with a new order of atoms. It is the same order as putting both $-CH_3$ groups on carbon number two.

```
         H                                   H
         |                                   |
      H—C—H                              H—C—H
         |                                   |
   H     |   H   H                   H   H   |   H
   |     |   |   |                   |   |   |   |
H—C—C—C—C—H              H—C—C—C—C—H
  1|   2|  3|  4|                   4|  3|  2|  1|
   H     |   H   H                   H   H   |   H
         |                                   |
      H—C—H                              H—C—H
         |                                   |
         H            same as                H
```

We can make no more different structural isomers with a four-carbon chain and two $-CH_3$ groups. Take another carbon from the chain.

```
    H   H   H        H        H        H
    |   |   |        |        |        |
H—C—C—C—    —C—    —C—    —C—H.
   1|  2|  3|       4|       5|       6|
    H   H   H        H        H        H
```

With a three-carbon chain and three $-CH_3$ groups, there are no new structural isomers. All the combinations we can make have the same order as ones we have already made. Thus there are five structural isomers of C_6H_{14}.

7. You can name alkanes by applying the rules in Section 20.4 of your textbook.

   a. The longest continuous chain of carbon atoms has six carbons, so the parent alkane name is hexane.

$$\boxed{CH_3—CH—CH—CH_2—CH—CH_3}$$

$$\quad\;\; CH_3\;\; CH_3 \qquad\quad CH_3$$

   We need to number the parent chain, assigning number one to the end that is closest to the first branch. In this case, the first branch, a $–CH_3$ group, would be on carbon two regardless of which end we begin numbering from. But beginning the numbering from the left would locate the next branch, also a $–CH_3$, on carbon three. If we began from the right, the second substituent would be on carbon four.

$$\boxed{\overset{1}{CH_3}—\overset{2}{CH}—\overset{3}{CH}—\overset{4}{CH_2}—\overset{5}{CH}—\overset{6}{CH_3}}$$

$$\quad\;\; CH_3\;\; CH_3 \qquad\quad CH_3$$

   We now need to name and number each group on the main chain. All three branches are $–CH_3$ groups. These are equivalent to methane molecules with a hydrogen removed and are called methyl groups. There is a methyl group on carbons two, three, and five. When there is more than one of a kind of substituent, use the prefix di, tri, and so on to indicate the number of times the substituent appears. These substituents would be named 2,3,5-trimethyl. Because there is only one kind of substituent, we do not have to list them in alphabetical order. The entire name is 2,3,5-trimethylhexane.

   b. The longest continuous chain of carbon atoms has five carbons, so the parent alkane name is pentane.

$$\boxed{CH_3—CH_2—CH—CH_2—CH_3}$$

$$CH—CH_3$$

$$|$$

$$CH_3$$

   We can number this alkane from either end. There is only one substituent, and it is on the middle carbon.

$$\boxed{\overset{1}{CH_3}—\overset{2}{CH_2}—\overset{3}{CH}—\overset{4}{CH_2}—\overset{5}{CH_3}}$$

$$CH—CH_3$$

$$|$$

$$CH_3$$

   The group on carbon three is an isopropyl group, so this molecule is called 3-isopropylpentane.

c. The longest continuous chain of carbon atoms has five carbon atoms, so the parent alkane name is pentane. Remember that the longest chain will not always be drawn horizontally on the page.

$$CH_3-CH-CH-CH_3$$
$$\quad\quad CH_2 \quad CH_3$$
$$\quad\quad CH_3$$

Number the chain beginning with the end on the right side in order to give the first substituent the lowest possible number.

$$CH_3-\overset{3}{CH}-\overset{2}{CH}-\overset{1}{CH_3}$$
$$\quad\quad {}^{4}CH_2 \quad CH_3$$
$$\quad\quad {}^{5}CH_3$$

There are two substituents, both methyl groups, on carbons two and three. The substituents are named 2,3-dimethyl. The whole name is 2,3-dimethylpentane.

d. The longest continuous chain of carbon atoms has seven carbons, so the parent alkane name is heptane.

$$\quad\quad\quad\quad CH_3$$
$$\quad\quad\quad\quad CH_2$$
$$CH_3-CH_2-CH-CH-CH-CH_2-CH_3$$
$$\quad CH_2 \quad CH_3$$
$$\quad CH_3$$

You can also draw the box another way, which gives a chain of equal length.

$$\quad\quad\quad\quad {}^{1}CH_3$$
$$\quad\quad\quad\quad {}^{2}CH_2$$
$$CH_3-CH_2-\overset{5}{CH}-\overset{4}{CH}-\overset{3}{CH}-CH_2-CH_3$$
$$\quad {}^{6}CH_2 \quad CH_3$$
$$\quad {}^{7}CH_3$$

Number the chain from either end.

```
                              CH3
                              |
                              CH2
  1      2      3      4      5|     6      7
 CH3—CH2—CH—CH—CH—CH2—CH3
               |      |
              CH2   CH3
               |
              CH3
```

The substituents are a $-CH_3$, or methyl group, on carbon four, and two CH_3CH_2-, or ethyl groups, on carbons three and five. The two ethyl groups are named 3,5-diethyl. When assembling the substituent names, list ethyl before methyl because the groups must be listed alphabetically. The group names are 3,5-diethyl-4-methyl. The entire name is 3,5-diethyl-4-methylheptane.

e. The longest continuous chain of carbons has three carbons, so the parent alkane name is propane.

```
        CH3
        |
 CH3—C—CH3
        |
        CH3
```

Number the chain from either end.

```
        CH3
  1    2|    3
 CH3—C—CH3
        |
        CH3
```

The substituents are two methyl groups on carbon number two. The substituents are named 2,2-dimethyl. The entire name is 2,2-dimethylpropane.

8. One way to tell whether or not two molecules are structural isomers is to name them. If two molecules have the same name, they are not structural isomers. They are the same molecule. If two molecules have the same number of carbons and hydrogens, and they have different names, they are structural isomers. Molecule **a** has five carbons in the longest chain, and molecule **b** also has five carbons in the longest chain.

```
 CH3—CH—CH—CH3                    CH3
      |     |                       |
     CH3  CH2—CH3            CH3—CH   CH3
                                    |     |
                                   CH2—CH
                                          |
                                         CH3

         a                            b
```

Number both molecules from the left. Each molecule has two substituents, two methyl groups. The substituents on molecule **a** are 2,3-dimethyl, and on molecule **b** they are 2,4-dimethyl. The entire name for molecule **a** is 2,3-dimethylpentane, and for molecule **b** it is 2,4-dimethylpentane. The names are different. Each molecule has five carbons and twelve hydrogens, so they are structural isomers.

9. Writing structural formulas from names is the reverse of writing names from formulas. First find the parent alkane name, and write a carbon skeleton with the same number of carbons. Number the chain from either direction. Determine how many substituents there are and where they are attached; then put them on the chain. Fill in hydrogens so that each carbon is surrounded by four other atoms.

a. 2-methyl-5-*sec*-butyloctane has the parent alkane name octane.

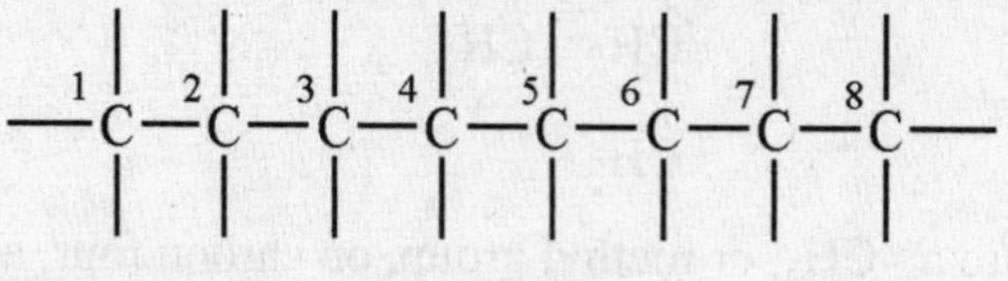

—C[1]—C[2]—C[3]—C[4]—C[5]—C[6]—C[7]—C[8]—

2-methyl indicates that there is a methyl group on carbon two.

—C[1]—C[2](—CH_3)—C[3]—C[4]—C[5]—C[6]—C[7]—C[8]—

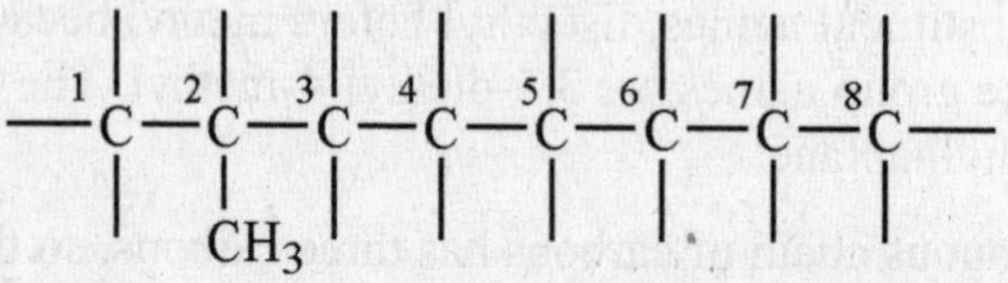

4-*sec*-butyl indicates that there is a *sec*-butyl group on carbon four.

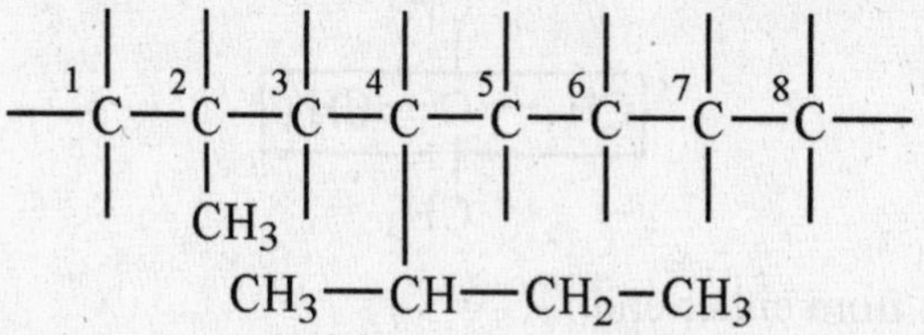

—C[1]—C[2](—CH_3)—C[3]—C[4](—CH_3—CH—CH_2—CH_3)—C[5]—C[6]—C[7]—C[8]—

Now fill in the hydrogens so that each carbon is surrounded by four atoms.

CH_3—CH(CH_3)—CH_2—CH(CH_3—CH—CH_2—CH_3)—CH_2—CH_2—CH_2—CH_3

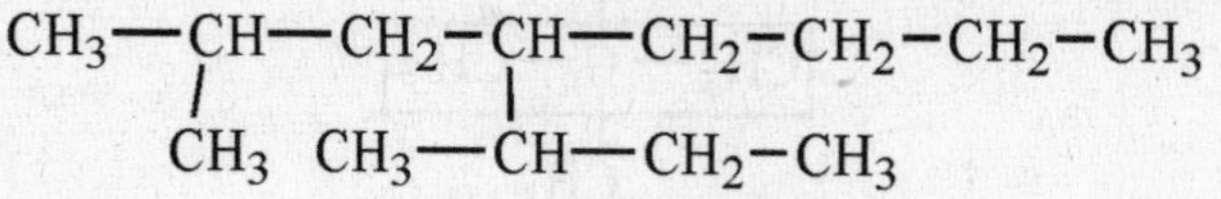

b. 2-methylpropane has the parent alkane name propane.

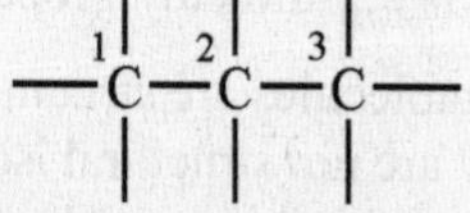

—C[1]—C[2]—C[3]—

2-methyl indicates a methyl group on carbon two.

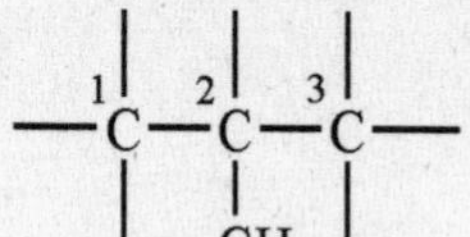

—C[1]—C[2](—CH_3)—C[3]—

Now fill in the hydrogens so that each carbon is surrounded by four atoms.

CH_3—CH(CH_3)—CH_3

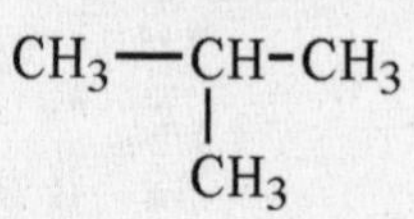

c. 3-ethylpentane has the parent alkane name pentane.

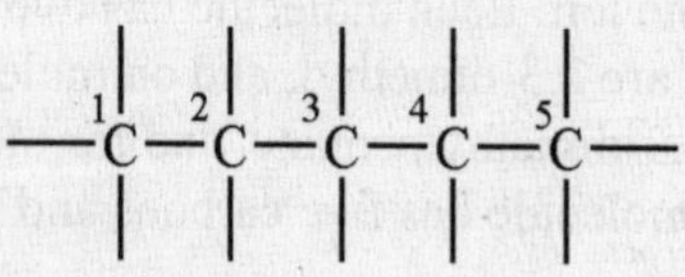

—C[1]—C[2]—C[3]—C[4]—C[5]—

3-ethyl indicates an ethyl group on carbon three.

$$\overset{1}{C}-\overset{2}{C}-\underset{\underset{CH_3}{|}}{\underset{CH_2}{\overset{3}{C}}}-\overset{4}{C}-\overset{5}{C}$$

Now fill in the hydrogens so that each carbon is surrounded by four atoms.

$$CH_3-CH_2-\underset{\underset{CH_3}{|}}{\underset{CH_2}{CH}}-CH_2-CH_3$$

d. 2,2,4-trimethylhexane has the parent alkane name hexane.

$$\overset{1}{C}-\overset{2}{C}-\overset{3}{C}-\overset{4}{C}-\overset{5}{C}-\overset{6}{C}$$

2,2,4-trimethyl indicates there are three methyl groups, two on carbon two and one on carbon four.

$$\overset{1}{C}-\underset{CH_3}{\overset{CH_3}{\overset{2}{C}}}-\overset{3}{C}-\underset{CH_3}{\overset{4}{C}}-\overset{5}{C}-\overset{6}{C}$$

Now, fill in the hydrogen atoms so that each carbon is surrounded by four atoms.

$$CH_3-\underset{CH_3}{\overset{CH_3}{C}}-CH_2-\underset{CH_3}{CH}-CH_2-CH_3$$

10.

a. 2-ethylbutane has the structure

$$CH_3-\underset{\underset{CH_3}{|}}{\underset{CH_2}{CH}}-CH_2-CH_3$$

The longest carbon chain has not four, but five carbons, so the correct alkane name is pentane. There is a methyl group on carbon three, so this is 3-methylpentane.

b. 1-methylpentane has the structure

$$\begin{array}{l} \overset{2}{C}H_2-\overset{3}{C}H_2-\overset{4}{C}H_2-\overset{5}{C}H_2-\overset{6}{C}H_3 \\ | \\ \overset{1}{C}H_3 \end{array}$$

The longest chain has six carbons, not five, so the correct name is hexane.

c. 4,4-dimethylhexane has the structure

$$\overset{1}{C}H_2-\overset{2}{C}H_2-\overset{3}{C}H_2-\overset{4}{C}(CH_3)_2-\overset{5}{C}H_2-\overset{6}{C}H_3$$

The longest chain has six carbons, so this is a hexane. The two methyl groups are on carbon four. Numbering the chain from the other end would put the methyl groups on carbon three. We want to number from the end that gives the substituents the lowest numbers, so 3,3-dimethylhexane would be the correct name.

d. 2-ethyl-2-methylheptane has the structure

$$\begin{array}{l} \quad\quad\;\; CH_3 \\ \quad\quad\;\; | \\ CH_3-\overset{3}{C}-\overset{4}{C}H_2-\overset{5}{C}H_2-\overset{6}{C}H_2-\overset{7}{C}H_2-\overset{8}{C}H_3 \\ \quad\quad\;\; | \\ \quad\quad\;\; \overset{2}{C}H_2 \\ \quad\quad\;\; | \\ \quad\quad\;\; \overset{1}{C}H_3 \end{array}$$

The longest chain has eight carbons, not seven, so the correct alkane name is octane. There are two methyl groups on carbon three, so this is 3,3-dimethyloctane.

11. Petroleum is a mixture of hydrocarbons containing molecules with various numbers of carbons in the chain. Usually, as the number of carbons in an alkane chain increases, the boiling point increases. Petroleum can be separated into different fractions by boiling.

12. First draw the structure of pentane.

$$\overset{1}{C}H_3-\overset{2}{C}H_2-\overset{3}{C}H_2-\overset{4}{C}H_2-\overset{5}{C}H_3$$

Then, begin at one end and remove a hydrogen atom and replace it with a chlorine atom.

$$\begin{array}{l} \overset{1}{C}H_2-\overset{2}{C}H_2-\overset{3}{C}H_2-\overset{4}{C}H_2-\overset{5}{C}H_3 \\ | \\ Cl \end{array}$$

There is only one structural isomer that can be produced from pentane by removing a hydrogen from carbon one. The order of bonding does not change by replacing either of the other two hydrogen atoms on carbon number one.

$$\overset{1}{CH_2}(Cl)-\overset{2}{CH_2}-\overset{3}{CH_2}-\overset{4}{CH_2}-\overset{5}{CH_3}$$

same as $Cl-\overset{1}{CH_2}-\overset{2}{CH_2}-\overset{3}{CH_2}-\overset{4}{CH_2}-\overset{5}{CH_3}$

same as $\overset{1}{CH_2}(Cl)-\overset{2}{CH_2}-\overset{3}{CH_2}-\overset{4}{CH_2}-\overset{5}{CH_3}$ (Cl above carbon 1)

Now, remove a hydrogen atom from carbon two and replace it with a chlorine atom. There is one structural isomer with a chlorine atom on carbon two.

$$\overset{1}{CH_3}-\overset{2}{CH}(Cl)-\overset{3}{CH_2}-\overset{4}{CH_2}-\overset{5}{CH_3}$$

Remove a hydrogen atom from carbon three, and replace it with a chlorine atom. There is one structural isomer with a chlorine atom on carbon three.

$$\overset{1}{CH_3}-\overset{2}{CH_2}-\overset{3}{CH}(Cl)-\overset{4}{CH_2}-\overset{5}{CH_3}$$

Substituting a chlorine atom on carbon atoms four or five does not produce a new structural isomer, so there are three structural isomers.

13.

a. $$CH_3-CH(CH_3)-CH_3 + Cl_2 \xrightarrow{h\nu} CH_3-CH(CH_3)-CH_2-Cl + HCl$$

In this reaction, a chlorine atom substitutes for a hydrogen atom.

b. $CH_4 + 2O_2 \longrightarrow CO_2 + 2H_2O$

This is an example of a combustion reaction.

c. $$CH_3-CH_2-CH_3 \xrightarrow[500°C]{Cr_2O_3} CH_3-CH=CH_2 + H_2$$

This is an example of a dehydrogenation reaction.

14. The rules for naming alkenes and alkynes are given in Section 20.7 of your textbook.

a. The longest chain that contains the double bond has five carbons.

$$\boxed{CH_3-CH_2-CH_2-C=CH_2}$$
$$\quad\quad\quad\quad\quad\quad\quad | $$
$$\quad\quad\quad\quad\quad\quad\quad CH_2-CH_3$$

(The CH_2CH_3 group is attached to the C of the double bond.)

Replace the -ane ending of pentane with -ene to produce pentene. Number the carbon chain from the end closest to the double bond.

$$\overset{5}{CH_3}-\overset{4}{CH_2}-\overset{3}{CH_2}-\overset{2}{C}(CH_2CH_3)=\overset{1}{CH_2}$$

Give the location of the double bond by putting a number in front of the alkene name. This molecule is 1-pentene. There is an ethyl group on carbon two, so this molecule would be named 2-ethyl-1-pentene.

b. The longest chain that contains the triple bond has seven carbons.

$$CH_3-CH(CH_3)-C\equiv C-CH_2-CH(CH_3)-CH_3$$

Replace the -ane ending of heptane with -yne to produce heptyne. Number the carbon chain from the end closest to the triple bond.

$$\overset{1}{CH_3}-\overset{2}{CH}(CH_3)-\overset{3}{C}\equiv\overset{4}{C}-\overset{5}{CH_2}-\overset{6}{CH}(CH_3)-\overset{7}{CH_3}$$

Give the location of the triple bond by putting a three in front of heptyne. There are two methyl groups attached to the chain on carbons two and six. This molecule would be named 2,6-dimethyl-3-heptyne.

c. The longest chain that contains the double bond has four carbons.

$$CH_3-C(CH_3)=C(CH_3)-CH_3$$

Replace the -ane ending of butane with -ene to produce butene. Number the carbon chain from either end to give the location of the double bond the smallest number.

$$\overset{1}{CH_3}-\overset{2}{C}(CH_3)=\overset{3}{C}(CH_3)-\overset{4}{CH_3}$$

Give the location of the double bond by putting a number in front of the alkene name. This molecule is 2-butene. There are two methyl groups attached to the main chain on carbons two and three. This molecule would be named 2,3-dimethyl-2-butene.

15.

a. $$CH_3-C(CH_3)=C(CH_3)-CH_3 + H_2 \xrightarrow{\text{catalyst}} CH_3-CH(CH_3)-CH(CH_3)-CH_3$$

b. $$CH_3-CH_2-CH_2-CH=CH_2 + Cl_2 \longrightarrow CH_3-CH_2-CH_2-CH(Cl)-CH_2Cl$$

c.

$$CH_3-CH=\underset{\underset{CH_3}{|}}{C}-CH_2-CH_3 + Br_2 \longrightarrow CH_3-\underset{\underset{Br}{|}}{CH}-\overset{\overset{Br}{|}}{\underset{\underset{CH_3}{|}}{C}}-CH_2-CH_3$$

16.

a. This molecule is usually called toluene.

b. This molecule is named as a pentane with two substituents. There is a chlorine atom on carbon three and a phenyl group on carbon two. The name of this molecule is 3-chloro-2-phenylpentane.

$$\overset{1}{CH_3}-\overset{2}{CH}-\overset{3}{\underset{\underset{Cl}{|}}{CH}}-\overset{4}{CH_2}-\overset{5}{CH_3}$$

(phenyl group on carbon 2)

c. This molecule is chlorobenzene.

d. This molecule is nitrobenzene.

17.

a. (benzene ring with NO_2 on carbons 1 and 2; ring positions numbered 1–6) is 1,2-dinitrobenzene or o-dinitrobenzene

b. (benzene ring with OH on carbon 1 and Cl on carbon 4; ring positions numbered 1–6) is 4-chlorophenol or p-chlorophenol

c. (benzene ring with CH_3 on carbon 1 and CH_2-CH_3 on carbon 4; ring positions numbered 1–6) is 4-ethyltoluene or p-ethyltoluene

d. (structure: benzene ring with CH_3 at carbon 1 and Br at carbon 3; ring carbons numbered 1–6) is 3-bromotoluene or m-bromotoluene

18.

a. The longest chain that contains the –OH has five carbons.

$$\overset{1}{CH_3}-\overset{2}{C}(CH_3)_2-\overset{3}{CH}(OH)-\overset{4}{CH_2}-\overset{5}{CH_3}$$

Number the chain from the left side so that the –OH group and the substituents have the lowest possible numbers. Drop the -e ending from pentane, and replace with -ol to produce pentanol. Locate the –OH group with a number. This molecule would be 3-pentanol. There are two methyl groups on carbon two, so the entire name would be 2,2-dimethyl-3-pentanol. The carbon to which the –OH is attached is bonded to two hydrocarbon fragments, so this is a secondary alcohol.

b. The longest chain that contains the –OH has four carbons.

$$\overset{1}{CH_3}-\overset{2}{C}(CH_3)(OH)-\overset{3}{CH_2}-\overset{4}{CH_3}$$

Number the chain from the top so that the –OH group has the lowest possible number. Drop the -e ending from butane, and add -ol to produce butanol. Locate the –OH group with a number. This molecule would be 2-butanol. There is a methyl group on carbon two, so the entire name is 2-methyl-2-butanol. The carbon to which the –OH is attached is bonded to three hydrocarbon fragments, so this is a tertiary alcohol.

c. The longest chain that contains the –OH group has six carbons.

$$\overset{6}{CH_3}-\overset{5}{CH_2}-\overset{4}{C}(CH_3)(Br)-\overset{3}{CH}(CH_3)-\overset{2}{CH_2}-\overset{1}{CH_2}-OH$$

Begin numbering from the right to give the –OH the lowest possible number. Drop the -e ending of hexane and add -ol to produce hexanol. Locate the –OH group with a number. This molecule would be 1-hexanol. There are two methyl groups on carbons three and four, and a bromo group on carbon four. The entire name would be 4-bromo-3,4-dimethyl-l-hexanol. The carbon to which the –OH is attached is bonded to one hydrocarbon fragment, so this is a primary alcohol.

d. The longest chain that contains the –OH has eight carbons.

$$\begin{array}{l} \qquad\quad CH_3-CH-CH_3 \\ \qquad\qquad\quad | \\ CH_3-\overset{2}{CH}-\overset{3}{CH_2}-\overset{4}{CH}-\overset{5}{CH_2}-\overset{6}{CH_2}-\overset{7}{CH_2}-\overset{8}{CH_3} \\ \qquad\ | \\ \quad {}_1\,CH_2-OH \end{array}$$

Begin numbering from the carbon that has the –OH. Drop the -e ending of octane, and add -ol to produce octanol. Locate the –OH group with a number. This molecule would be 1-octanol. There is a methyl group on carbon two and an isopropyl group on carbon four. The entire name would be 2-methyl-4-isopropyl-l-octanol. The carbon to which the –OH is attached is bonded to one hydrocarbon fragment, so this is a primary alcohol.

e. The longest chain has one carbon. Assign this carbon number one.

$$CH_3OH$$

Drop the -e ending of methane and add -ol to produce methanol. Because there is only one carbon, we do not need a number to locate the –OH group. It can be found only on carbon one. There are no substituents, so methanol is the entire name. This molecule is also known by the common names methyl alcohol and wood alcohol. The carbon to which the –OH is attached is bonded to one hydrocarbon fragment, so this is a primary alcohol.

f. The longest chain that contains the –OH has nine carbons.

$$\begin{array}{l} CH_3-\overset{7}{CH_2}-\overset{8}{CH}-\overset{9}{CH_2}-CH_3 \\ \qquad\qquad\quad | \\ \qquad\quad {}_6\,CH_2-\overset{5}{CH_2}-\overset{4}{CH}-\overset{3}{CH_2}-\overset{2}{CH}-\overset{1}{CH_3} \\ \qquad\qquad\qquad\qquad\quad |\qquad\qquad\ \ | \\ \qquad\qquad\qquad\qquad CH_2\qquad\quad OH \\ \qquad\qquad\qquad\qquad\quad | \\ \qquad\qquad\qquad\qquad CH_3 \end{array}$$

Begin numbering from the right to give the –OH the lowest possible number. Drop the -e ending of nonane, and add -ol to produce nonanol. Locate the –OH with a number. This molecule would be 2-nonanol. There are two ethyl groups on carbons four and seven. This molecule would be named 4,7-diethyl-2-nonanol. The carbon to which the –OH is attached is bonded to two hydrocarbon fragments, so this is a secondary alcohol.

19. Match the alcohols below with the appropriate description.

a. methanol	commonly known as wood alcohol
b. phenol	used in the production of plastics
c. ethanol	found in alcoholic beverages
d. ethylene glycol	used in antifreeze

20.

a. $C_6H_5-CH_2-OH \xrightarrow{\text{oxidation}} C_6H_5-\overset{O}{\overset{\|}{C}}-H$

b.

$$CH_3-\underset{\displaystyle CH_3}{\underset{|}{CH}}-CH_2-OH \xrightarrow{\text{oxidation}} CH_3-\underset{\displaystyle CH_3}{\underset{|}{CH}}-C(=O)H$$

c.

$$CH_3-CH_2-\underset{\displaystyle OH}{\underset{|}{CH}}-CH_2-CH_3 \xrightarrow{\text{oxidation}} CH_3-CH_2-\underset{\displaystyle O}{\underset{||}{C}}-CH_2-CH_3$$

21.

a. The longest chain that contains the carbonyl has four carbons. Begin numbering from the end that has the aldehyde group.

$$\overset{4}{CH_3}-\overset{3}{\underset{\displaystyle CH_3}{\overset{\displaystyle CH_3}{C}}}-\overset{2}{\underset{\displaystyle CH_3}{CH}}-\overset{1}{C}(=O)H$$

Drop the -e ending of butane, and add -al to produce butanal. Since the carbonyl in aldehydes is always found on an end, do not use a number to locate the carbonyl. There are three methyl groups, two on carbon three and one on carbon two. This molecule would be named 2,3,3-trimethylbutanal.

b. The longest chain that contains the carbonyl has six carbons. Begin numbering from the end that is closest to the ketone group.

$$\underset{1}{CH_3}-\underset{2}{\overset{\displaystyle O}{\overset{||}{C}}}-\underset{3}{CH_2}-\underset{4}{\overset{\displaystyle CH_2-CH_3}{CH}}-\underset{5}{CH_2}-\underset{6}{CH_3}$$

Drop the -e of hexane, and add -one to produce hexanone. Locate the ketone carbonyl with a number, 2-hexanone. There is an ethyl group on carbon four, so the entire name would be 4-ethyl-2-hexanone.

c. The longest chain that contains the carbonyl has seven carbons. Begin numbering from the end that has the aldehyde group.

$$H(O=)\overset{1}{C}-\underset{\displaystyle Br}{\overset{2}{CH}}-\overset{3}{CH_2}-\overset{4}{CH_2}-\overset{5}{CH_2}-\overset{6}{CH_2}-\overset{7}{CH_3}$$

Drop the -e ending of heptane, and add -al to produce heptanal. There is a bromo group on carbon two, so the entire name is 2-bromoheptanal.

d. The longest chain that contains the carbonyl has six carbons. Begin numbering from the right to give the ketone the lowest number.

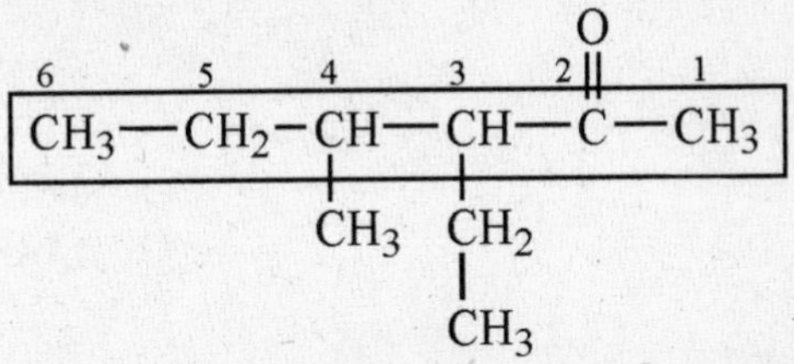

$$\overset{6}{CH_3}-\overset{5}{CH_2}-\overset{4}{CH}(CH_3)-\overset{3}{CH}(CH_2CH_3)-\overset{2}{C}(=O)-\overset{1}{CH_3}$$

Drop the -e ending of hexane, and add -one to produce hexanone. Locate the ketone with a number. This molecule would be 2-hexanone. There is an ethyl group on carbon three and a methyl group on carbon four. The entire name would be 3-ethyl-4-methyl-2-hexanone.

e. The longest chain that contains the carbonyl has four carbons. Begin numbering from the right to give the ketone carbonyl the lowest number.

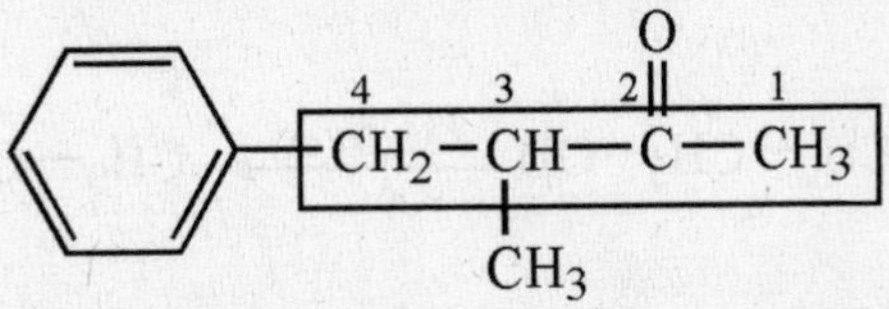

$$C_6H_5-\overset{4}{CH_2}-\overset{3}{CH}(CH_3)-\overset{2}{C}(=O)-\overset{1}{CH_3}$$

Drop the -e of butane, and add -one to produce butanone. Locate the ketone with a number. This molecule would be 2-butanone. There is a methyl group on carbon three and a phenyl group on carbon four. The name would be 3-methyl-4-phenyl-2-butanone.

f. The longest chain that contains the carbonyl has five carbons. Begin numbering from the end with the aldehyde group.

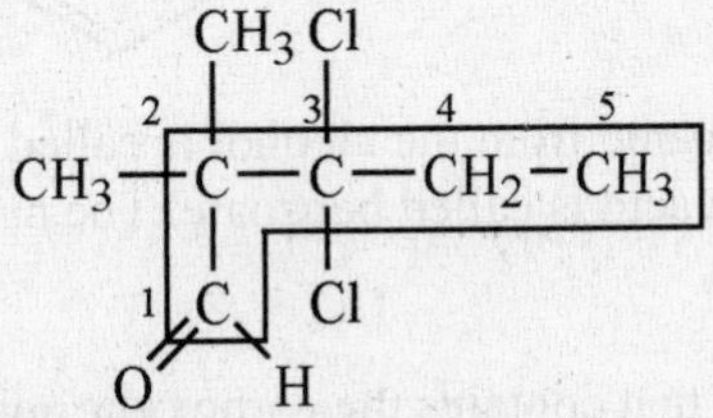

$$CH_3-\overset{2}{C}(CH_3)(\overset{1}{C}HO)-\overset{3}{C}(Cl)_2-\overset{4}{CH_2}-\overset{5}{CH_3}$$

Drop the -e of pentane and add -al to produce pentanal. The aldehyde group does not need a number because it is always on the end. There are two methyl groups on carbon two and two chloro groups on carbon three. The name would be 3,3-dichloro-2,2-dimethylpentanal.

22.

a. 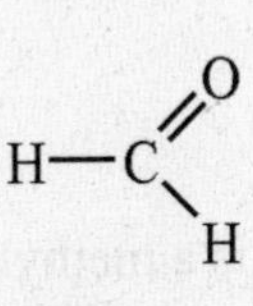 $H-C(=O)-H$

aldehyde

b. 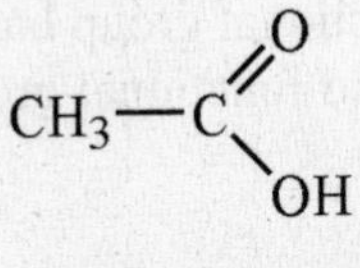 $CH_3-C(=O)-OH$

carboxylic acid

c. CH_3-CH_2-OH

alcohol

d. 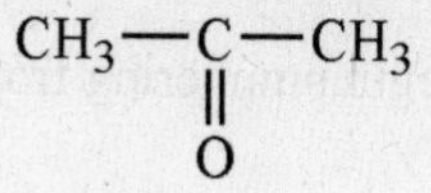 $CH_3-C(=O)-CH_3$

ketone

e. 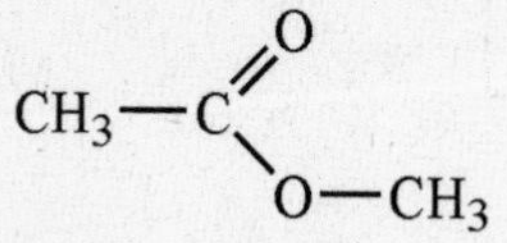 $CH_3-C(=O)-O-CH_3$

ester

23.

a. 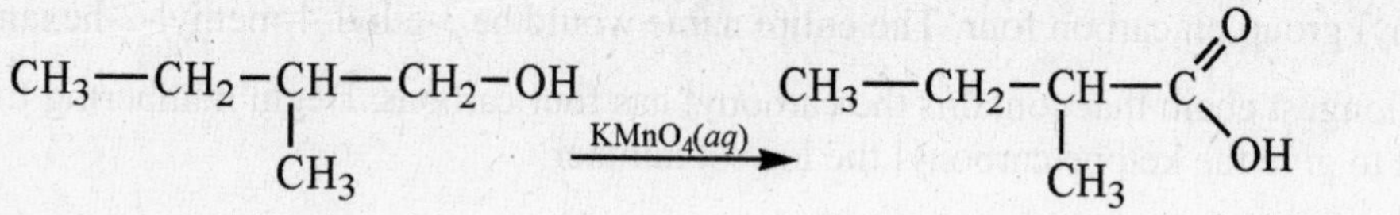 $CH_3-CH_2-CH(CH_3)-CH_2-OH \xrightarrow{KMnO_4(aq)} CH_3-CH_2-CH(CH_3)-C(=O)OH$

b. 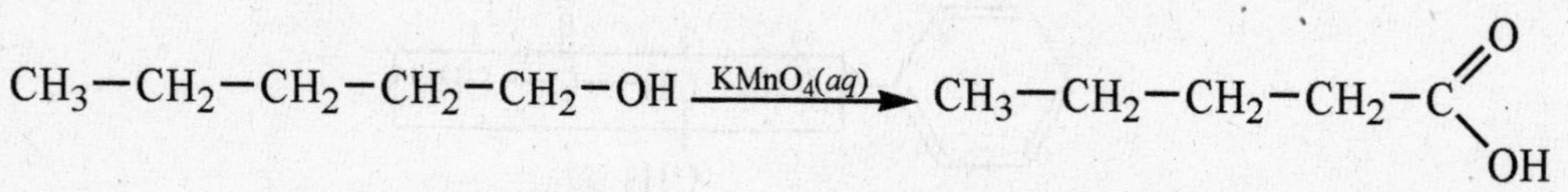 $CH_3-CH_2-CH_2-CH_2-CH_2-OH \xrightarrow{KMnO_4(aq)} CH_3-CH_2-CH_2-CH_2-C(=O)OH$

24.

a. 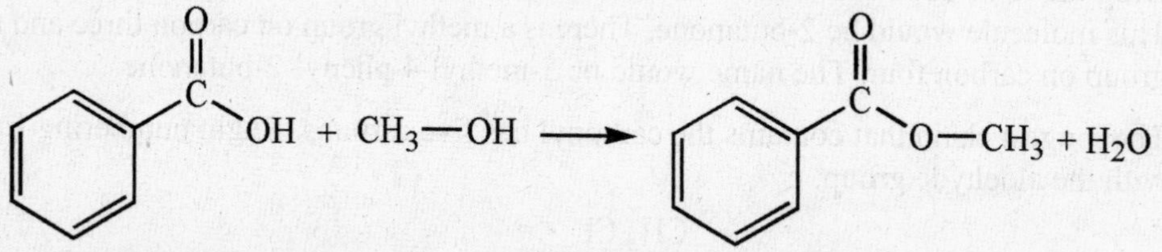 $C_6H_5-C(=O)-OH + CH_3-OH \longrightarrow C_6H_5-C(=O)-O-CH_3 + H_2O$

b. The part of the ester that came from the alcohol is called methyl. The part of the ester that came from the carboxylic acid is called benzoate. The name is methylbenzoate.

25.

a. The longest carbon chain that contains the carboxylic acid functional group has seven carbons. Number the chain from the end that has the carboxylic acid group.

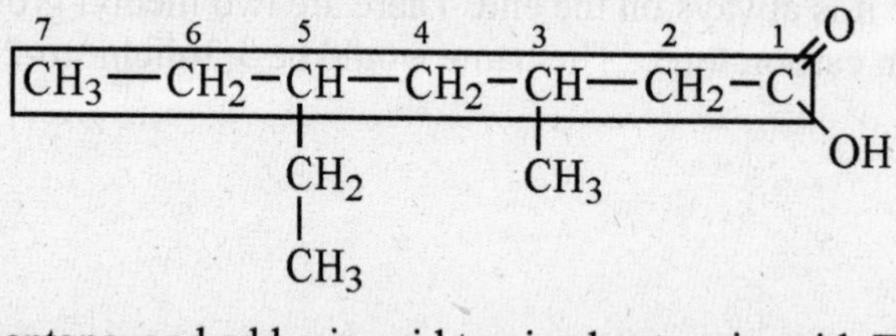

$\overset{7}{CH_3}-\overset{6}{CH_2}-\overset{5}{CH}(CH_2CH_3)-\overset{4}{CH_2}-\overset{3}{CH}(CH_3)-\overset{2}{CH_2}-\overset{1}{C}(=O)OH$

Drop the -e of heptane, and add -oic acid to give heptanoic acid. There is a methyl group on carbon three and an ethyl group on carbon five. The name is 5-ethyl-3-methylheptanoic acid.

b. The longest chain of carbons that contains the carboxylic acid functional group has four carbons. Begin numbering from the right to give the carboxylic acid functional group the lowest number.

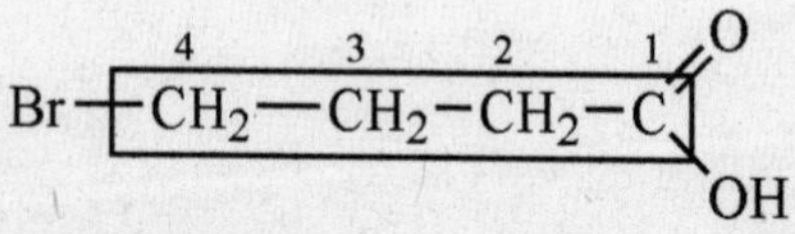

$Br-\overset{4}{CH_2}-\overset{3}{CH_2}-\overset{2}{CH_2}-\overset{1}{C}(=O)OH$

Drop the -e ending and add -oic acid to give butanoic acid. There is a bromo group on carbon four, so the name would be 4-bromobutanoic acid.

c. The longest chain of carbons that contains the carboxylic acid functional group has four carbons. Begin numbering from the right to give the carboxylic acid functional group the lowest number.

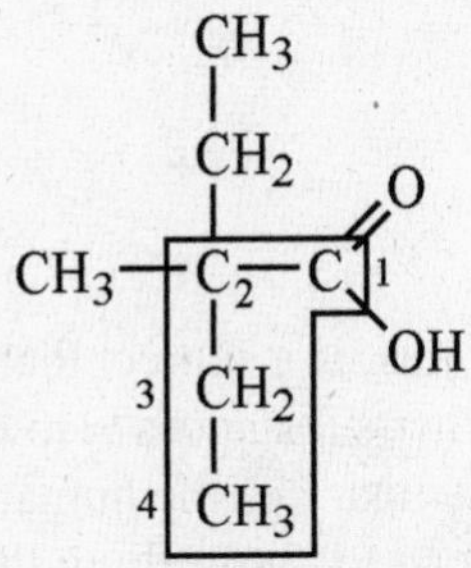

$$CH_3-\overset{\overset{CH_3}{|}\atop\overset{CH_2}{|}}{\underset{\underset{4\ CH_3}{|}}{\underset{3\ CH_2}{\underset{|}{C_2}}}}-\overset{O}{\overset{\|}{C_1}}-OH$$

Drop the -e ending, and add -oic acid to give butanoic acid. There is a methyl group and an ethyl group on carbon two. The entire name would be 2-ethyl-2-methylbutanoic acid.

26.

a. 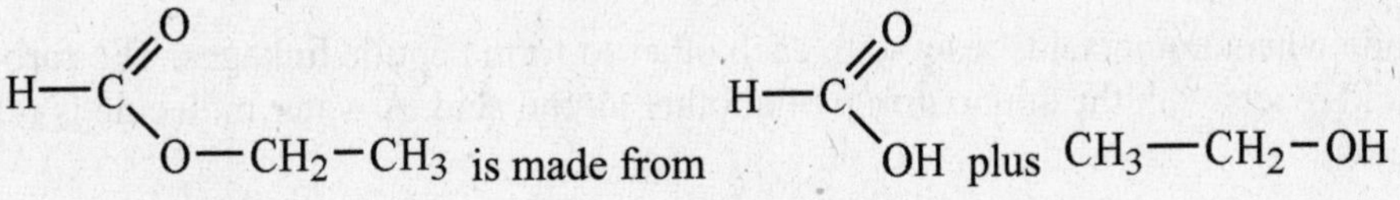 $H-\overset{O}{\overset{\|}{C}}-O-CH_2-CH_3$ is made from $H-\overset{O}{\overset{\|}{C}}-OH$ plus CH_3-CH_2-OH

b. 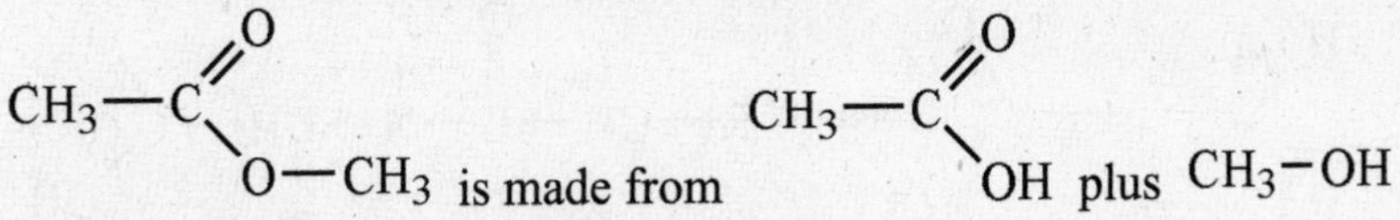 $CH_3-\overset{O}{\overset{\|}{C}}-O-CH_3$ is made from $CH_3-\overset{O}{\overset{\|}{C}}-OH$ plus CH_3-OH

27. The polymer would have the structure below.

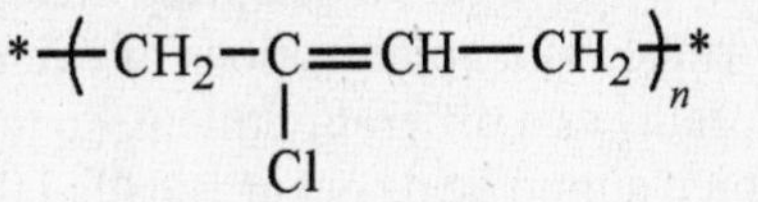

$$*\!\left(CH_2-\underset{\underset{Cl}{|}}{C}=CH-CH_2\right)_n\!*$$

CHAPTER 21

Biochemistry

INTRODUCTION

Biochemistry, the chemistry of living systems, is a subject that concerns everyone. How do our bodies extract chemical energy from sugar and other substances? How can we find cures for diseases? The answers to these questions lie in the biochemistry of the human body. To begin to understand how the complex biochemical systems work, we need to know about the kinds of molecules that are important in living organisms. They are often large organic molecules, and they contain atoms and functional groups with characteristics with which we are already familiar.

CHAPTER DISCUSSION

Proteins are made when amino acids react with each other to form peptide linkages. The carboxyl group of one amino acid reacts with the amino group of another amino acid. A water molecule is removed during the process.

$$\mathrm{R{-}CH(NH_2){-}C({=}O){-}\boxed{OH \quad H}{-}N(H){-}CH(R'){-}C({=}O){-}OH \longrightarrow R{-}CH(NH_2){-}C({=}O){-}N(H){-}CH(R'){-}C({=}O){-}OH + H_2O}$$

The bond formed between two amino acids is called a peptide linkage. There is an almost endless number of proteins that could be formed from combinations of twenty different amino acids. Each different sequence of amino acids produces a different protein, so the order in which the amino acids occur is important. The order in which amino acids occur is called the primary structure of a protein.

The secondary structure of a protein is the shape the protein chain assumes. Two common secondary structures are the alpha-helix and the pleated sheet.

The tertiary structure of a protein is the three-dimensional shape of the protein molecule. Some proteins are globular in shape, some are elongated. A protein can have several areas of alpha-helix that are separated from each other by bends. The protein folds at each bend, giving the molecule its tertiary structure.

Carbohydrates are a large class of biomolecules composed mainly of carbon, hydrogen, and oxygen. They are a major source of food and serve as structural components in plants. Some carbohydrates are simple sugars, or monosaccharides (such as glucose), some are disaccharides (such as sucrose or table sugar), and some are polysaccharides (such as starch and cellulose).

Nucleic acids store the information necessary for life to continue from generation to generation. The nucleic acid that stores genetic information and transmits information from one generation to the next is deoxyribonucleic acid, or DNA. Another nucleic acid that helps translate the genetic information into proteins is ribonucleic acid, or RNA. Both DNA and RNA are composed of smaller building blocks called nucleotides. A nucleotide is made from three parts, a five-carbon monosaccharide, a nitrogen-containing organic base, and a phosphate group.

LEARNING REVIEW

1. What functions do carbon, phosphorus and magnesium perform in the human body?
2. What are two major types of proteins, and how do their functions differ?
3. Which of the amino acids below have hydrophilic and which have hydrophobic side chains?

   a. $HO-C_6H_4-CH(NH_2)-C(=O)OH$

   b. $HO-C(=O)-CH_2-CH(NH_2)-C(=O)OH$

   c. $H_3C-CH(CH_3)-CH_2-CH(NH_2)-C(=O)OH$

4. What is the structure of the dipeptide made from the two amino acids below?

   $HO-C_6H_4-CH_2-CH(NH_2)-C(=O)OH \; + \; HS-CH_2-CH(NH_2)-C(=O)OH \longrightarrow$

5. Show the sequences of all tripeptides that can be made from the amino acids phenylalanine (phe), glycine (gly), proline (pro), and aspartic acid (asp).
6. Explain the difference between the primary, secondary and tertiary structures of proteins.
7. Two types of secondary structures are the α-helix and the pleated sheet. What are the major characteristics of each?
8. Which amino acid plays a special role in maintaining the tertiary structure of proteins?
9. Use a lock and key analogy to explain how an enzyme can catalyze a reaction.
10. What two functional groups are characteristic of monosaccharides?
11. Draw the structure of a tetrose that has an aldehyde functional group.
12. The disaccharide sucrose, or table sugar, is made from which two monosaccharides?
13. What difference between starch and cellulose causes starch to be a food source for humans while cellulose is not?
14.
    a. What are the parts of a nucleotide?
    b. How do the nucleotides of RNA differ from nucleotides of DNA?
15. Show the structure of a nucleotide.
16. Show why the bases cytosine and guanine and the bases adenine and thymine pair with each other in DNA.
17. How is it thought that DNA reproduces itself?

18. How are proteins synthesized from the DNA code?

19.

    a. Draw the structure of the fat made from a molecule of glycerol and three molecules of oleic acid.

    b. Would this fat likely be a solid or a liquid at room temperature?

20. Explain how soap cleans away greasy dirt.

21. To form wax molecules, molecules with which two functional groups must combine?

22. The steroids are a diverse group of molecules. What structural feature do they all have in common?

ANSWERS TO LEARNING REVIEW

1. Carbon is the backbone of all organic molecules in the body. Phosphorus is present in cell membranes and plays an important role in energy transfer in cells. Magnesium is required for proper functioning of some enzymes.

2. The two major types are fibrous and globular. Fibrous proteins provide structure and shape while globular proteins perform chemical work.

3.

    a. The side chain on tyrosine is a benzene ring that has an –OH substituted for one of the hydrogen atoms. The –OH bond is polar, just as the –OH bond in water is. The polarity of the –OH bond makes the side chain of tyrosine hydrophilic.

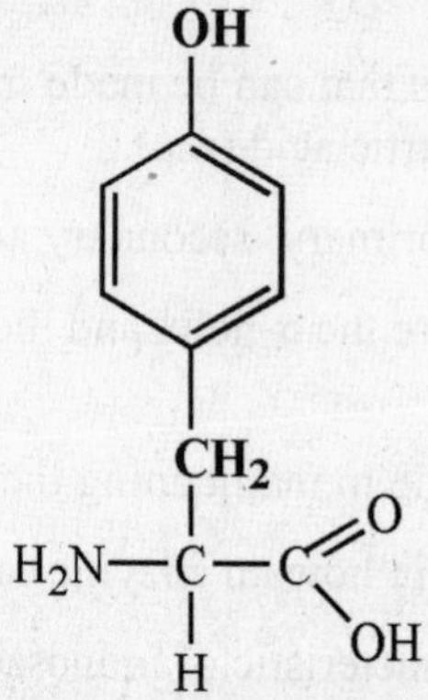

OH

CH_2

H_2N—C—C=O

H OH

    b. Aspartic acid has a side chain that has a $-CH_2$ and a –COOH. The –OH bond is polar, and so is the –CO bond. The polarity of the –COOH makes the side chain of aspartic acid hydrophilic.

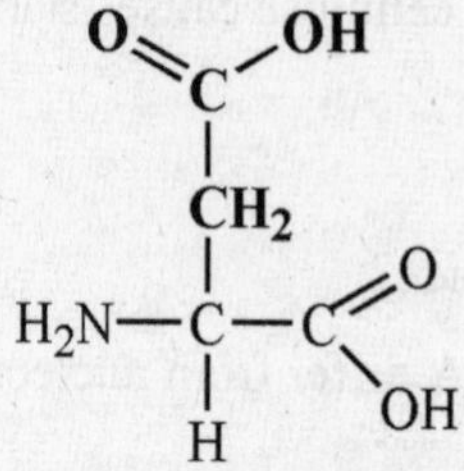

O=C—OH

CH_2

H_2N—C—C=O

H OH

c. Leucine has a side chain composed entirely of carbon and hydrogen. Carbon-hydrogen bonds are not polar, so the side chain of leucine is hydrophobic.

$$H_3C-CH(CH_3)-CH_2-C(H)(NH_2)-C(=O)OH$$

4.

$$HO-C_6H_4-CH_2-CH(NH_2)-C(=O)OH + HS-CH_2-CH(NH_2)-C(=O)OH \rightarrow$$

$$HO-C_6H_4-CH_2-CH(NH_2)-C(=O)-NH-CH(CH_2SH)-C(=O)OH$$

5. To help you determine all the sequences, begin with one of the amino acids, and write all the possible sequences beginning with this one amino acid. Then choose another amino acid, and write all the sequences that begin with the second amino acid. Continue until you have written sequences that begin with each of the amino acids given.

phe-gly-pro	gly-phe-pro	pro-gly-phe	asp-phe-pro
phe-gly-asp	gly-phe-asp	pro-gly-asp	asp-phe-gly
phe-pro-gly	gly-pro-phe	pro-phe-asp	asp-pro-phe
phe-pro-asp	gly-pro-asp	pro-phe-gly	asp-pro-gly
phe-asp-pro	gly-asp-pro	pro-asp-phe	asp-gly-phe
phe-asp-gly	gly-asp-phe	pro-asp-gly	asp-gly-pro

6. The primary structure of proteins is the amino acid sequence. The secondary structure of proteins is the arrangement of the protein chain or the arrangement of the amino acids to form an α-helix or a pleated sheet. How the α-helices or pleated sheets are bent in relation to one another is the tertiary structure of a protein.

7. In the α-helix the amino acids are coiled like a spiral staircase. This makes a long region of protein. Areas of protein with lots of α-helix are springy and elastic. In a pleated sheet, protein chains are bent back one or more times to form a sheet of protein chains. Pleated sheets are strong and resistant to stretching.

8. The amino acid cysteine helps proteins maintain their unique tertiary structures. The –SH side chains of two cysteine molecules can react to form a disulfide linkage that holds the protein chain in a fixed tertiary structure.

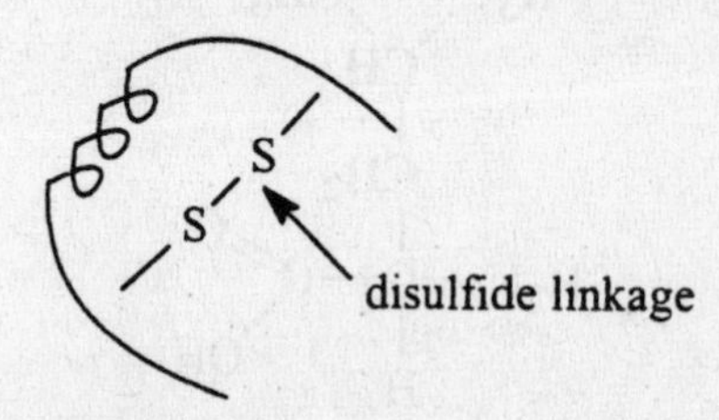

9. For some enzymes it is believed that the enzyme has a shape that fits the shape of the substrate, just like a key fits a lock. When the enzyme and the substrate join together, a reaction occurs. The enzyme then releases the product that has formed.

10. Monosaccharides all have hydroxyl and carbonyl functional groups. The carbonyl can be either an aldehyde or a ketone.

11. There are several tetroses with an aldehyde group. The structure of one is presented below.

H—C=O
H—C—OH
H—C—OH
CH_2OH

12. Sucrose is made from the monosaccharides glucose and fructose.

13. Both starch and cellulose are made from glucose molecules. The way the glucose molecules are joined together in starch is different from the way they are joined in cellulose. Enzymes in our bodies can break the links between glucose molecules in starch, but not in cellulose.

14.

a. All nucleotides are made from a phosphate group, a nitrogen-containing organic base, and a five-carbon sugar.

b. In RNA, the sugar is ribose while in DNA the sugar is deoxyribose. Some of the organic bases also differ between DNA and RNA. DNA and RNA have cytosine, adenine and guanine. Thymine is found only in DNA, and uracil is found only in RNA.

15.

NH_2
N N
N N
OH
HO—P—O—CH_2 C O C H H H H C—C OH OH
O
phosphate
ribose
organic base adenine

16. Cytosine and guanine pair with each other in DNA because the hydrogen-bonding sites on the molecule are complementary. The three hydrogen bonds between the two molecules hold cytosine and guanine together. Adenine and thymine molecules on complementary DNA strands are also held together by hydrogen bonds. Two hydrogen bonds form between adenine and thymine molecules.

17. There is evidence to show that two complementary DNA chains unwind and that new strands are made by pairing bases along the old chains. The end result is two new chains, each complementary to the old one.

18. DNA stores the information required to produce proteins needed by organisms. A length of the DNA chain called a gene contains the information for one protein. The DNA transmits the information it contains by synthesizing a chain of RNA called messenger RNA (mRNA). Once the mRNA has been produced, it moves away from DNA to the site of protein synthesis. The mRNA and small bodies called ribosomes begin the synthesis of protein. Another kind of RNA, called transfer RNA (tRNA), brings up amino acids one by one to join them on the end of the growing protein chain. The mRNA contains the information that determines which amino acids join, and in what order.

19.

a.

$$
\begin{array}{l}
CH_2-O-\overset{\overset{\large O}{\|}}{C}-(CH_2)_7-CH{=}CH-(CH_2)_7CH_3 \\
\ |\\
CH-O-\overset{\overset{\large O}{\|}}{C}-(CH_2)_7-CH{=}CH-(CH_2)_7CH_3 \\
\ |\\
CH_2-O-\overset{\overset{\large O}{\|}}{C}-(CH_2)_7-CH{=}CH-(CH_2)_7CH_3
\end{array}
$$

b. Fats made of unsaturated fatty acids are often liquids at room temperature. This fat would likely be a liquid at room temperature because oleic acid is an unsaturated fatty acid.

20. Soap molecules are the anions of long chain fatty acids. Each soap molecule has two parts. The carboxylate head has a negative charge and is attracted to polar water molecules. The remainder of the molecule is a long hydrocarbon tail that is not attracted to water. The hydrocarbon tail is hydrophobic. When soap is added to water, the hydrocarbon tail does not want to associate with the polar water molecules, so tails from many soap molecules associate together to form a soap micelle in which the polar heads face outward toward the water.

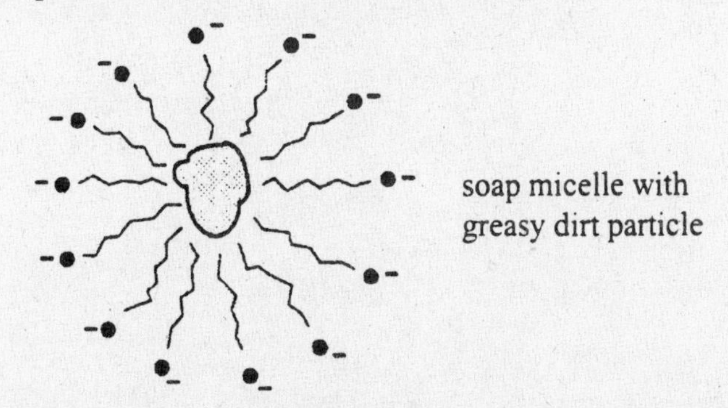

soap micelle with greasy dirt particle

Much of the dirt we wish to remove is greasy dirt. This kind of dirt is not washed away by water because it is hydrophobic. When soap micelles come in contact with greasy dirt, the dirt is lifted from the surface and enters the inside of the micelle. The hydrophobic dirt would rather be in contact with the hydrophobic hydrocarbon tails than with water. The soap micelles that contain the greasy dirt are washed away with water.

21. Waxes are esters with long carbon chains. Esters are made from a carboxylic acid and an alcohol.

22. All steroids have a basic ring structure called the steroid nucleus. Each steroid has different substituents on the rings.

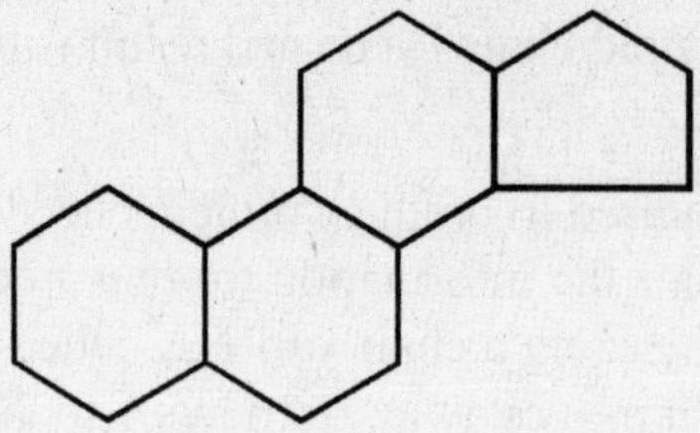